SpringerBriefs in Quantitative Finance

For further volumes:
http://www.springer.com/series/8784

L. C. G. Rogers

Optimal Investment

Springer

L. C. G. Rogers
Statistical Laboratory
University of Cambridge
Cambridge
UK

ISSN 2192-7006 ISSN 2192-7014 (electronic)
ISBN 978-3-642-35201-0 ISBN 978-3-642-35202-7 (eBook)
DOI 10.1007/978-3-642-35202-7
Springer Heidelberg New York Dordrecht London

Library of Congress Control Number: 2012953459

Mathematics Subject Classification (2010): 91G10, 91G70, 91G80, 49L20, 65K15

JEL Classifications: G11, C61, D53, D90

Printed on acid-free paper

Springer is part of Springer Science+Business Media (www.springer.com)

For Judy, Ben, and Stefan

Preface

Whether you work in fund management, a business school, or a university economics or mathematics department, the title of this book, *Optimal Investment*, promises to be of interest to you. Yet its contents are, I guess, not as you would expect. Is it about the practical issues of portfolio selection in the real world? No; though it does not ignore those issues. Is it a theoretical treatment? Yes; though often issues of completeness and rigour are suppressed to allow for a more engaging account. The general plan of the book is to set out the most basic problem in continuous-time portfolio selection, due in its original form to Robert Merton. The first chapter presents this problem and some variants, along with a range of methods that can be used for its solution, and the treatment here is quite careful and thorough. There is even a complete verification of the solution of the Merton problem! But the theorem/proof style of academic mathematical finance quickly palls, and anyone with a lively imagination will find this too slow-moving to hold the attention.[1] So in the second chapter, we allow ourselves to run ahead of proof, and present a large number of quite concrete and fascinating examples, all inspired by the basic Merton problem, which rested on some overly specific assumptions. We ask what happens if we take the Merton problem, and change the assumptions in various ways: How does the solution change if there are transaction costs? If the agent's preferences are different? If the agent is subject to various kinds of constraint? If the agent is uncertain about model parameters? If the underlying asset dynamics are more general? This is a chapter of variations on the basic theme, and many of the individual topics could be, have been, or will be turned into full-scale academic papers, with a lengthy literature survey, a careful specification of all the spaces in which the processes and variables take values, a detailed and thorough verification proof, maybe even some study of data to explore how well the new story accounts from some phenomenon. Indeed, this is very much the pattern of the subject, and is something I hope this book will help to put

[1] ... but anyone who wants to get to grips with the details will find exemplary presentations in [30] or [21], for example.

in its proper place. Once the reader has finished with Chapter 2, it should be abundantly clear that in all of these examples we can very quickly write down the equations governing the solution; we can very rarely solve them in closed form; so at that point we either have to stop or do some numerics. What remains constitutes the conventional steps of a formal academic dance. So the treatment of the examples emphasizes the essentials—the formulation of the equations for the solution, any reduction or analysis which can make them easier to tackle, and then numerically calculating the answer so that we can *see* what features it has—and leaves the rest for later. There follows a brief chapter discussing numerical methods for solving the problems. There is likely little here that would surprise an expert in numerical analysis, but discussions with colleagues would indicate that the Hamilton-Jacobi-Bellman equations of stochastic optimal control are perhaps not as extensively studied within PDE as other important areas. And the final chapter takes a look at some actual data, and tries to assess just how useful the preceding chapters may be in practice.

As with most books, there are many people to thank for providing support, encouragement, and guilt. Much of the material herein has been given as a graduate course in Cambridge for a number of years, and each year by about the third lecture of the course students will come up to me afterwards and ask whether there is any book that deals with the material of the course—we all know what that signifies. At last I will be able to answer cheerfully and confidently that there is indeed a book which follows closely the content and style of the lectures! But this book would not have happened were it not for the invitations to give various short courses over the years: I am more grateful than I can say to Damir Filipovic; Anton Bovier; Tom Hurd and Matheus Grasselli; Masaaki Kijima, Yukio Muromachi, Hidetaka Nakaoka, and Keiichi Tanaka; and Ralf Korn for the opportunities their invitations gave me to spend time thinking through the problems explained in this book. I am indebted to Arieh Iserles who kindly provided me with numerous comments on the chapter on numerical methods; and I am likewise most grateful to my students over the years for their inputs and comments on various versions of the course, which have greatly improved what follows. And last but not least it is a pleasure to thank my colleagues at Cantab Capital Partners for allowing me to come and find out what the issues in fund management really are, and why none of what you will read in this book will actually help you if that is your goal.

Cambridge, October 2012 Chris Rogers

Contents

1 The Merton Problem 1

1.1 Introduction 1

1.2 The Value Function Approach 4

1.3 The Dual Value Function Approach 11

1.4 The Static Programming Approach 14

1.5 The Pontryagin-Lagrange Approach 17

1.6 When is the Merton Problem Well Posed? 20

1.7 Linking Optimal Solutions to the State-Price Density 22

1.8 Dynamic Stochastic General Equilibrium Models 23

1.9 CRRA Utility and Efficiency 28

2 Variations 29

2.1 The Finite-Horizon Merton Problem 30

2.2 Interest-Rate Risk 31

2.3 A Habit Formation Model 33

2.4 Transaction Costs 36

2.5 Optimisation under Drawdown Constraints 39

2.6 Annual Tax Accounting 43

2.7 History-Dependent Preferences 45

2.8 Non-CRRA Utilities 47

2.9 An Insurance Example with Choice of Premium Level 49

2.10 Markov-Modulated Asset Dynamics 53

2.11 Random Lifetime 57

2.12 Random Growth Rate 59

2.13 Utility from Wealth and Consumption 61

2.14 Wealth Preservation Constraint 62

2.15 Constraint on Drawdown of Consumption 64

2.16 Option to Stop Early 68

2.17 Optimization under Expected Shortfall Constraint 70

2.18 Recursive Utility 72

2.19 Keeping up with the Jones's . . . 73
2.20 Performance Relative to a Benchmark . . . 75
2.21 Utility from Slice of the Cake . . . 76
2.22 Investment Penalized by Riskiness . . . 77
2.23 Lower Bound for Utility . . . 79
2.24 Production and Consumption . . . 81
2.25 Preferences with Limited Look-Ahead . . . 84
2.26 Investing in an Asset with Stochastic Volatility . . . 88
2.27 Varying Growth Rate . . . 91
2.28 Beating a Benchmark . . . 94
2.29 Leverage Bound on the Portfolio . . . 96
2.30 Soft Wealth Drawdown . . . 97
2.31 Investment with Retirement . . . 99
2.32 Parameter Uncertainty . . . 102
2.33 Robust Optimization . . . 106
2.34 Labour Income . . . 110

3 Numerical Solution . . . 115
3.1 Policy Improvement . . . 117
3.1.1 Optimal Stopping . . . 120
3.2 One-Dimensional Elliptic Problems . . . 121
3.3 Multi-Dimensional Elliptic Problems . . . 123
3.4 Parabolic Problems . . . 127
3.5 Boundary Conditions . . . 130
3.6 Iterative Solutions of PDEs . . . 133
3.6.1 Policy Improvement . . . 133
3.6.2 Value Recursion . . . 134
3.6.3 Newton's Method . . . 134

4 How Well Does It Work? . . . 137
4.1 Stylized Facts About Asset Returns . . . 139
4.2 Estimation of μ: The 20s Example . . . 144
4.3 Estimation of V . . . 146

References . . . 151

Index . . . 153

Chapter 1
The Merton Problem

Abstract The first chapter of the book introduces the classical Merton problems of optimal investment over a finite horizon to maximize expected utility of terminal wealth; and of optimal investment over an infinite horizon to maximize expected integrated utility of running consumption. The workhorse method is to find the Hamilton-Jacobi-Bellman equations for the value function and then to try to solve these in some way. However, in a complete market we can often use the budget constraint as the necessary and sufficient restriction on possible consumption streams to arrive quickly at optimal solutions. The third main method is to use the Pontryagin-Lagrange approach, which is an example of dual methods.

1.1 Introduction

The story to be told in this book is in the style of a musical theme-and-variations; the main theme is stated, and then a sequence of variations is played, bearing more or less resemblance to the main theme, yet always derived from it. For us, the theme is the Merton problem, to be presented in this chapter, and the variations will follow in the next chapter.

What is the Merton problem? I use the title loosely to describe a collection of stochastic optimal control problems first analyzed by Merton [28]. The common theme is of an agent investing in one or more risky assets so as to optimize some objective. We can characterise the dynamics of the agent's wealth through the equation[1]

$$\begin{aligned} dw_t &= r_t w_t\, dt + n_t \cdot (dS_t - r_t S_t dt + \delta_t dt) + e_t\, dt - c_t\, dt && (1.1)\\ &= r_t (w_t - n_t \cdot S_t) dt + n_t \cdot (dS_t + \delta_t dt) + e_t\, dt - c_t\, dt. && (1.2) \end{aligned}$$

[1] Commonly, some of the terms of the wealth equation may be missing; we often assume $e \equiv 0$, and sometimes $\delta \equiv 0$.

L. C. G. Rogers, *Optimal Investment*, SpringerBriefs in Quantitative Finance,
DOI: 10.1007/978-3-642-35202-7_1, © Springer-Verlag Berlin Heidelberg 2013

for some given initial wealth w_0. In this equation, the *asset price process* S is a d-dimensional semimartingale, the *portfolio process* n is a d-dimensional previsible process, and the *dividend process* δ is a d-dimensional adapted process.[2] The adapted scalar processes e and c are respectively an endowment stream, and a consumption stream.The process r is an adapted scalar process, interpreted as the riskless rate of interest. The processes δ, S, r and e will generally be assumed given, as will the initial wealth w_0, and the agent must choose the portfolio process n and the consumption process c.

To explain a little how the wealth equation (1.1) arises, think what would happen if you invested nothing in the risky assets, that is, $n \equiv 0$; your wealth, invested in a bank account, would grow at the riskless rate r, with addition of your endowment e and withdrawal of your consumption c. If you chose to hold a fixed number $n_t = n_0$ of units of the risky assets, then your wealth w_t at time t would be made up of the market values $n_0^i S_t^i$ of your holding of asset i, $i = 1, \dots, d$, together with the cash you hold in the bank, equal to $w_t - n_0 \cdot S_t$. The cash in the bank is growing at rate r—which explains the first term on the right in (1.2)—and the ownership of n_0^i units of asset i delivers you a stream $n_0^i \delta_t^i$ of dividends.

Next, if you were to follow a piecewise constant investment strategy, where you just change your portfolio in a non-anticipating way at a finite set of stopping times, then the evolution between change times is just as we have explained it; at change times, the new portfolio you choose has to be funded from your existing resources, so there is no jump in your wealth. Thus we see that the evolution (1.1) is correct for any (left-continuous, adapted) piecewise constant portfolio process n, and by extension for any previsible portfolio process.

If we allow completely arbitrary previsible n, we immediately run into absurdities. For this reason, we usually restrict attention to portfolio processes n and consumption processes c such that the pair (n, c) is *admissible*.

Definition 1.1 The pair $(n_t, c_t)_{t \geq 0}$ is said to be *admissible for initial wealth* w_0 if the wealth process w_t given by (1.1) remains non-negative at all times. We use the notation

$$\mathscr{A}(w_0) \equiv \{(n, c) : (n, c) \text{ is admissible from initial wealth } w_0\} \tag{1.3}$$

We shall write $\mathscr{A} \equiv \cup_{w>0} \mathscr{A}(w)$ for the set of all admissible pairs (n, c).

Notational convention. *The portfolio held by an investor is sometimes characterized by the* number *of units of the assets held, sometimes by the* cash values *invested in the different assets. Depending on the particular context, either one may be preferable. As a notational convention, we shall always write n for a* **n***umber of assets,* and θ *for what the holding of assets is wor***th**.[3] *Thus if at time t we hold* n_t^i *units of asset i, whose time-t price is* S_t^i*, then we have the obvious identity*

[2] The notation $a \cdot b$ for $a, b \in \mathbb{R}^d$ denotes the scalar product of the two vectors.

[3] ... since the Greek letter θ corresponds to the English 'th'.

$$\theta_t^i = n_t^i S_t^i.$$

From time to time, it will be useful to characterize the portfolio held in terms of the proportion of wealth assigned to each of the assets. For this, we will let π_t^i *be the* **p***roportion of wealth invested in asset i at time t, so that the notations*

$$\theta_t^i = n_t^i S_t^i = \pi_t^i w_t$$

all give the same thing, namely, the cash value of the holding of asset i at time t.

We have discussed the controls available to the investor; what about his objective? Most commonly, we suppose that the agent is trying to choose (n, c) so as to obtain

$$\sup_{(n,c)\in\mathscr{A}(w_0)} E\left[\int_0^T u(t, c_t)\, dt + u(T, w_T)\right]. \tag{1.4}$$

The function u is supposed to be concave increasing in its second argument, and measurable in its first. The time horizon T is generally taken to be a positive constant. Special cases of this objective include

$$\sup_{(n,c)\in\mathscr{A}(w_0)} E\left[\int_0^\infty u(t, c_t)\, dt\right], \tag{1.5}$$

the *infinite-horizon problem*, and

$$\sup_{(n,0)\in\mathscr{A}(w_0)} E[u(w_T)], \tag{1.6}$$

the *terminal wealth problem*.

This then is the problem: the agent aims to achieve (1.4) when his control variables must be chosen so that the wealth process w generated by (1.1) remains non-negative. How shall this be solved? We shall see a variety of methods, but there is a very important principle underlying many of the approaches, worth explaining on its own.

Theorem 1.1 *(The Davis-Varaiya Martingale Principle of Optimal Control). Suppose that the objective is* (1.4)*, and that there exists a function* $V : [0, T] \times \mathbb{R}^+ \to \mathbb{R}$ *which is* $C^{1,2}$*, such that* $V(T, \cdot) = u(T, \cdot)$*. Suppose also that for any* $(n, c) \in \mathscr{A}(w_0)$

$$Y_t \equiv V(t, w_t) + \int_0^t u(s, c_s)\, ds \quad \text{is a supermartingale,} \tag{1.7}$$

and that for some $(n^*, c^*) \in \mathscr{A}(w_0)$ *the process Y is a martingale. Then* (n^*, c^*) *is optimal, and the value of the problem starting from initial wealth* w_0 *is*

$$V(0, w_0) = \sup_{(n,c)\in\mathscr{A}(w_0)} E\left[\int_0^T u(t, c_t)\, dt + u(T, w_T)\right]. \tag{1.8}$$

Proof From the supermartingale property of Y, we have for any $(n, c) \in \mathscr{A}(w_0)$

$$Y_0 = V(0, w_0) \geq E[Y_T] = E\left[\int_0^T u(t, c_t)\, dt + u(T, w_T)\right], \tag{1.9}$$

using the fact that $V(T, \cdot) = u(T, \cdot)$. Thus for any admissible strategy the value is no greater than $V(0, w_0)$; when we use (n^*, c^*), the value is equal to $V(0, w_0)$ since the (supermartingale) inequality in (1.9) becomes an equality. Hence (n^*, c^*) is optimal. □

Remarks (i) The Martingale Principle of Optimal Control (MPOC for short) is a very simple yet very useful result. It cannot be applied without care or thought, but is remarkably effective in leading us to the right answer, even if we have to resort to other methods to prove that it is the right answer.
(ii) Notice from the linearity of the wealth equation (1.1) that if $(n, c) \in \mathscr{A}(w)$ and $(n', c') \in \mathscr{A}(w')$, then $(pn + (1-p)n', pc + (1-p)c') \in \mathscr{A}(pw + (1-p)w')$ for any $p \in (0, 1)$. From the concavity of U we deduce immediately the following little result.

Proposition 1.1 *The value function $V(t, w)$ is concave increasing in its second argument.*

1.2 The Value Function Approach

The most classical methodology for solving a stochastic optimal control problem is the value function approach, and this is based on the MPOC. First we make the asset dynamics a bit more explicit; we shall suppose that

$$dS_t^i = S_t^i\Big(\sum_{j=1}^N \sigma^{ij} dW_t^j + \mu^i dt\Big), \tag{1.10}$$

where the σ^{ij} and the μ^i are constants, and W is an d-dimensional Brownian motion. We shall also suppose that the riskless rate of interest r is constant, and that the endowment process e and dividend process δ are identically zero. We express the equation (1.10) more compactly as

$$dS_t = S_t(\sigma \cdot dW + \mu dt). \tag{1.11}$$

Notice that the wealth equation (1.1) can be equivalently (and more usefully) expressed as

$$dw_t = rw_t\, dt + \theta_t \cdot (\sigma dW_t + (\mu - r)\, dt) - c_t\, dt. \tag{1.12}$$

How would we find some function V satisfying the hypotheses of Theorem (1.1)? The direct approach is just to write down the process Y from (1.7) and perform an Itô expansion, assuming that V possesses sufficient regularity:

$$\begin{aligned} dY_t &= V_t dt + V_w\, dw + \tfrac{1}{2} V_{ww}\, dw dw + u(t,c)\, dt \\ &= V_w \theta \cdot \sigma dW + \left\{ u(t,c) + V_t + V_w(rw + \theta \cdot (\mu - r) - c) + \tfrac{1}{2}|\sigma^T \theta|^2 V_{ww} \right\} dt. \end{aligned}$$

The stochastic integral term in the Itô expansion is a local martingale; if we could assume that it were a *martingale*, then the condition for Y to be a supermartingale whatever (θ, c) was in use would just be that the drift were non-positive. Moreover, if the supremum of the drift were equal to zero, then we should have that V was the value function, with the pointwise-optimizing (θ, c) constituting an optimal policy. Setting all the provisos aside for the moment, this would lead us to consider the equation

$$0 = \sup_{\theta, c} \left[u(t,c) + V_t + V_w(rw + \theta \cdot (\mu - r) - c) + \tfrac{1}{2}|\sigma^T \theta|^2 V_{ww} \right]. \tag{1.13}$$

This (non-linear) partial differential equation (PDE) for the unknown value function V is the *Hamilton-Jacobi-Bellman* (HJB) equation . If we have a problem with a finite horizon, then we shall have the boundary condition $V(T, \cdot) = u(T, \cdot)$; for an infinite-horizon problem, we do not have any boundary conditions to fix a solution, though in any given context, we may be able to deduce enough growth conditions to fix a solution. There are many substantial points where the line of argument just sketched meets difficulties:

1. Is there any solution to the PDE (1.13)?
2. If so, is there a unique solution satisfying boundary/growth conditions?
3. Is the supremum in (1.13) attained?
4. Is V actually the value function?

Despite this, for a given stochastic optimal control problem, writing down the HJB equation for that problem is usually a very good place to start. Why? The point is that *if we are able to find some V which solves the HJB equation*, then it is usually possible by direct means to verify that the V so found is actually the value function. If we are not able to find some V solving the HJB equation, then what have we actually achieved? Even if we could answer all the questions (1)–(4) above, all we know is that there is a value function and that it is the unique solution to (1.13); we do not know how the optimal policy (θ^*, c^*) looks, we do not know how the solution changes if we change any of the input parameters, in fact, we really cannot say anything interesting about the solution!!

The philosophy here is that we seek *concrete solutions* to optimal control problems, and general results on existence and uniqueness do not themselves help us to this goal. Usually, in order to get some reasonably explicit solution, we shall have to assume a simple form for the utility u, such as

$$u(t,x) = e^{-\rho t}\,\frac{x^{1-R}}{1-R}, \tag{1.14}$$

or

$$u(t,x) = -\gamma^{-1}\exp(-\rho t - \gamma x), \tag{1.15}$$

where $\rho, \gamma, R > 0$ and $R \neq 1$. Since the derivative of the utility (1.14) is just $u'(t,x) = e^{-\rho t}x^{-R}$, in the case $R = 1$ we understand it to be

$$u(t,x) = e^{-\rho t}\log(x). \tag{1.16}$$

All of these forms of the utility are very tractable, and if we do not assume one of these forms we will rarely be able to get very far with the solution.

Key example: the infinite-horizon Merton problem. To illustrate the main ideas in a simple and typical example, let's assume the constant-relative-risk-aversion (CRRA) form (1.14) for u, which we write as

$$u(t,x) \equiv e^{-\rho t}u(x) \equiv e^{-\rho t}\,\frac{x^{1-R}}{1-R}. \tag{1.17}$$

The aim is to solve the infinite-horizon problem; the agent's objective is to find the value function

$$V(w) = \sup_{(n,c)\in\mathscr{A}(w)} E\left[\int_0^\infty e^{-\rho t}\,\frac{c_t^{1-R}}{1-R}\,dt\right], \tag{1.18}$$

and the admissible (n, c) which attains the supremum, if possible. We shall see that this problem can be solved completely. The steps involved are:

- STEP 1: Use special features to guess the form of the solution;
- STEP 2: Use the HJB equation to find the solution;
- STEP 3: Find a simple bound for the value of the problem;
- STEP 4: Verify that the bound is attained for the conjectured optimal solution.

This strategy is applicable to many examples other than this one, and should be regarded as the main line of attack on a new problem. Let us see these steps played out.

STEP 1: USING SPECIAL FEATURES. What makes this problem easy is the fact that *because of scaling, we can write down the form of the solution*; indeed, we can immediately say that[4]

[4] We require of course that the problem is well-posed, that is, the supremum is finite. We shall have more to say on this in Section 1.6.

$$V(w) = \gamma_M^{-R} u(w) \equiv \gamma_M^{-R} \frac{w^{1-R}}{1-R} \tag{1.19}$$

for some constant $\gamma_M > 0$. Thus finding the solution to the optimal investment/consumption problem reduces to identifying the constant γ_M. Why do we know that V has this simple form?

Proposition 1.2 **(Scaling)** *Suppose that the problem* (1.18) *is well posed. Then the value function takes the form* (1.19).

Proof By the linearity of the wealth equation (1.12), it is clear that

$$(n, c) \in \mathscr{A}(w) \Leftrightarrow (\lambda n, \lambda c) \in \mathscr{A}(\lambda w)$$

for any $\lambda > 0$. Hence

$$\begin{aligned} V(\lambda w) &= \sup_{(n,c)\in\mathscr{A}(\lambda w)} E\left[\int_0^\infty e^{-\rho t} u(c_t)\, dt\right] \\ &= \sup_{(n,c)\in\mathscr{A}(w)} E\left[\int_0^\infty e^{-\rho t} u(\lambda c_t)\, dt\right] \\ &= \sup_{(n,c)\in\mathscr{A}(w)} \lambda^{1-R} E\left[\int_0^\infty e^{-\rho t} u(c_t)\, dt\right] \\ &= \lambda^{1-R} V(w). \end{aligned}$$

□

Taking $w = 1$ gives the result.

STEP 2: USING THE HJB EQUATION TO FIND THE VALUE. Can we go further, and actually identify the constant γ_M appearing in (1.19)? We certainly can, and as we do so we learn everything about the solution. If we consider

$$V(t, w) = \sup_{(n,c)\in\mathscr{A}(w)} E\left[\int_t^\infty e^{-\rho s} \frac{c_s^{1-R}}{1-R}\, ds \,\middle|\, w_t = w\right],$$

then it is clear from the time-homgeneity of the problem that

$$V(t, w) = e^{-\rho t} V(w), \tag{1.20}$$

where V is as defined at (1.18). In view of the scaling form (1.19) of the solution, we now suspect that

$$V(t, w) = e^{-\rho t} \gamma_M^{-R} u(w), \tag{1.21}$$

and we just have to identify the constant γ_M. For this, we return to the HJB Eq. (1.13). The HJB equation involves an optimization over θ and c, which can be performed explicitly.

OPTIMIZATION OVER θ. The optimization over θ is easy[5]:

$$(\sigma\sigma^T)\theta V_{ww} = -(\mu - r)V_w,$$

whence

$$\theta^* = -\frac{V_w}{V_{ww}}\,(\sigma\sigma^T)^{-1}(\mu - r). \tag{1.22}$$

Using the suspected form (1.21) of the solution, this is simply

$$\boxed{\theta^* = wR^{-1}\,(\sigma\sigma^T)^{-1}(\mu - r).} \tag{1.23}$$

To interpret this solution, let us introduce the notation

$$\boxed{\pi_M \equiv R^{-1}(\sigma\sigma^T)^{-1}(\mu - r),} \tag{1.24}$$

a constant N-vector, called the *Merton portfolio*. What (1.23) tells us is that for each i, and for all $t > 0$, the cash value of the optimal holding of asset i should be

$$(\theta_t^*)^i = w_t\,\pi_M^i;$$

so *the optimal investment in asset i is proportional to current wealth* w_t, with constant of proportionality π_M^i. Looking back, this form is hardly surprising, in view of the scaling property of the objective.

OPTIMIZATION OVER c. For the optimization over c, if we introduce the convex dual function

$$\tilde{u}(y) \equiv \sup_x\{u(x) - xy\} \tag{1.25}$$

of u, then we have for $u(x) = x^{1-R}/(1-R)$ that

$$\tilde{u}(y) = -\frac{y^{1-\tilde{R}}}{1-\tilde{R}}, \tag{1.26}$$

where $\tilde{R} = R^{-1}$. Thus the optimization over c develops as

$$\sup_c\{u(t,c) - cV_w\} = e^{-\rho t}\sup_c\{u(c) - ce^{\rho t}V_w\} = e^{-\rho t}\tilde{u}(e^{\rho t}V_w).$$

---

[5] Notice that the value function V should be concave in w, so V_{ww} will be negative.

Substituting in the suspected form (1.21) of the solution, this gives us

$$\sup_c\{u(t,c) - cV_w\} = e^{-\rho t}\tilde{u}((\gamma_M w)^{-R}) = -e^{-\rho t}\frac{(\gamma_M w)^{1-R}}{1-\tilde{R}} = e^{-\rho t}\,\frac{R}{1-R}\,(\gamma_M w)^{1-R},$$

with optimizing c^* proportional to w:

$$\boxed{c^* = \gamma_M w.} \tag{1.27}$$

Again, the fact that optimal consumption is proportional to wealth is not surprising in view of the scaling property of the objective.

PUTTING IT ALL TOGETHER. Returning the candidate value function (1.21) to the HJB Eq. (1.13), we find that

$$\begin{aligned}0 &= e^{-\rho t}\left[\frac{R}{1-R}\,(\gamma_M w)^{1-R} - \rho\gamma_M^{-R}u(w) + rw\gamma_M^{-R}w^{-R} + \tfrac{1}{2}\gamma_M^{-R}w^{1-R}|\kappa|^2/R\right] \\ &= \frac{e^{-\rho t}w^{1-R}\gamma_M^{-R}}{1-R}\left[R\gamma_M - \rho - (R-1)(r + \tfrac{1}{2}|\kappa|^2/R)\right]\end{aligned}$$

where

$$\kappa \equiv \sigma^{-1}(\mu - r) \tag{1.28}$$

is the *market price of risk* vector. This gives the value of γ_M:

$$\boxed{\gamma_M = R^{-1}\{\rho + (R-1)(r + \tfrac{1}{2}|\kappa|^2/R)\},} \tag{1.29}$$

and hence the value function of the Merton problem (see (1.21)), $V_M(w) \equiv V(t,w)$, as

$$\boxed{V_M(w) = \gamma_M^{-R}\,u(w).} \tag{1.30}$$

We now believe that we know the form of the optimal solution to the infinite-horizon Merton problem; *we invest proportionally to wealth* (1.23), *and consume proportionally to wealth* (1.27), *where the constants of proportionality are given by* (1.24) and (1.29) *respectively.*

FINISHING OFF. There are two issues to deal with; firstly, what happens if the expression (1.29) for γ_M is negative? Secondly, can we *prove* that the solution we have found actually is optimal?

The first of these questions relates to the question of whether or not the problem is ill-posed, and the answer has to be specific to the exact problem under consideration. The second question is actually much more general, and the method we use to deal with it applies in many examples. For this reason, we shall answer the second question first, assuming that γ_M given by (1.29) is positive, then return to the first question.

Suppose that the initial wealth w_0 is given, and consider the evolution of the wealth w^* under the conjectured optimal control; we see

$$\begin{aligned} dw_t^* &= w_t^* \left\{ \pi_M \cdot \sigma dW_t + (r + \pi_M \cdot (\mu - r) - \gamma_M) dt \right\} \\ &= w_t^* \left\{ R^{-1}\kappa \cdot dW_t + (r + R^{-1}|\kappa|^2 - \gamma_M) dt \right\} \end{aligned}$$

which is solved by

$$w_t^* = w_0 \exp\left[R^{-1}\kappa \cdot W_t + (r + \tfrac{1}{2}R^{-2}|\kappa|^2(2R-1) - \gamma_M)t\right] \tag{1.31}$$

STEP 3: FINDING A SIMPLE BOUND. The proof of optimality is based on the trivial inequality:

$$u(y) \le u(x) + (y - x)u'(x) \qquad (x, y > 0), \tag{1.32}$$

which expresses the geometrically obvious fact that the tangent to the concave function u at $x > 0$ lies everywhere above the graph of u. If we consider any admissible (n, c) then, we are able to bound the objective by

$$\begin{aligned} E\left[\int_0^\infty e^{-\rho t}\, u(c_t)\, dt\right] &\le E\left[\int_0^\infty e^{-\rho t}\, \left\{u(c_t^*) + (c_t - c_t^*)u'(c_t^*)\right\} dt\right] \\ &= E\int_0^\infty e^{-\rho t}\, u(c_t^*)\, dt + E\left[\int_0^\infty (c_t - c_t^*)\zeta_t\, dt\right], \end{aligned} \tag{1.33}$$

where we have abbreviated

$$\zeta_t \equiv e^{-\rho t}u'(c_t^*) \propto \exp(-\kappa \cdot W_t - (r + \tfrac{1}{2}|\kappa|^2)t) \tag{1.34}$$

after some simplifications using the explicit form of w^*. Now the key point is that ζ *is a state-price density*, also called a stochastic discount factor; we have the property that for any admissible (n, c)

$$Y_t \equiv \zeta_t w_t + \int_0^t \zeta_s c_s\, ds \quad \text{is a local martingale.} \tag{1.35}$$

This may be verified directly by Itô calculus from the wealth equation (1.1) in this example, and we leave it to the reader to carry out this check. In general, we expect that *the marginal utility of the optimal consumption should be a state-price density*, which will be explained (in a non-rigorous fashion) later.[6] The importance of the statement (1.35) is that since the wealth and consumption are non-negative, the process Y is in fact a non-negative supermartingale, and hence

---

[6] See Section 1.8.

$$w_0 = Y_0 \geq E[Y_\infty] \geq E\left[\int_0^\infty \zeta_s c_s \, ds\right]. \tag{1.36}$$

STEP 4: VERIFYING THE BOUND IS ATTAINED FOR THE CONJECTURED OPTIMUM. One last piece remains, and that is to verify the equality

$$w_0 = E\left[\int_0^\infty \zeta_s c_s^* \, ds\right] \tag{1.37}$$

for the optimal consumption process c^*, and again this can be established by direct calculation using the explicit form of c^*. Combining (1.33), (1.36) and (1.37) gives us finally that for any admissible (n, c)

$$E\left[\int_0^\infty e^{-\rho t} \, u(c_t) \, dt\right] \leq E\left[\int_0^\infty e^{-\rho t} \, u(c_t^*) \, dt\right], \tag{1.38}$$

which proves optimality of the conjectured optimal solution (n^*, c^*).

1.3 The Dual Value Function Approach

This approach should be regarded as a variant of the basic value function approach of Section 1.2, in that it offers a different way to tackle the HJB equation, but all the issues which arise there still have to be dealt with. We can expect this approach to simplify the HJB equation, but we will still have to go through the steps of verifying the solution; nevertheless, the simplifications resulting here are dramatic.

The basic idea is to take the HJB equation in its form (1.13) and transform it suitably. Notice that we will require the Markovian setup with constant σ, μ, and r, but we will not require any particular form for the utility U; all we ask is that it is concave strictly increasing in its second argument, and continuous in its first, and that

$$\lim_{x \to \infty} u'(t, x) = 0, \tag{1.39}$$

which is necessary for the optimization to be well posed.

Since we know that the value function is concave, the derivative V_w is monotone decreasing, so we are able to define a new coordinate system

$$(t, z) = (t, V_w(t, w)) \tag{1.40}$$

for (t, z) in $\mathrm{A} \equiv \{(t, z) : V_w(t, \infty) < z < V_w(t, 0)\}$. Now we define a function $J : \mathrm{A} \to \mathbb{R}$ by

$$J(t, z) = V(t, w) - wz, \tag{1.41}$$

and we notice that by standard calculus we have the relations

$$J_z = -w, \tag{1.42}$$
$$J_t = V_t, \tag{1.43}$$
$$J_{zz} = -1/V_{ww}. \tag{1.44}$$

Now when we take the HJB equation (1.13) and optimize over θ and c we obtain

$$0 = \tilde{u}(t, V_w) + V_t + rwV_w - \tfrac{1}{2}|\kappa|^2 \frac{V_w^2}{V_{ww}} \tag{1.45}$$
$$= \tilde{u}(t, z) + J_t - rzJ_z + \tfrac{1}{2}|\kappa|^2 z^2 J_{zz} \tag{1.46}$$

which is a *linear PDE for the unknown J*. Here, $\tilde{u}(t, z) \equiv \sup\{u(t, x) - zx\}$ is the convex dual of u.

The key example again. To see this in action, let us take the infinite-horizon Merton problem, and suppose that

$$u(t, x) = e^{-\rho t} u(x) \tag{1.47}$$

for some concave increasing non-positive[7] u, *which we do not assume has any particular form.* At this level of generality, the approach of Section 1.2 is invalidated, since it depended heavily on scaling properties which we do not now have. Nonetheless, the dual value function approach still works.

In this instance, we know[8] that $V(t, w) = e^{-\rho t} v(w)$ for some concave function v which is to be found. From the definition (1.41) of the dual value function J, we have

$$\begin{aligned} J(t, z) &= e^{-\rho t} v(w) - wz \\ &= e^{-\rho t}(v(w) - wze^{\rho t}) \\ &\equiv e^{-\rho t} j(ze^{\rho t}), \end{aligned}$$

say. Notice that since u is non-positive, it has to be that V is also non-positive, and that j is non-positive.

If we introduce the variable $y = ze^{\rho t}$, simple calculus gives

$$J_t = -\rho e^{-\rho t} j(y) + \rho z j'(y), \quad J_z = j'(y), \quad J_{zz} = e^{\rho t} j''(y)$$

and substituting into (1.46) gives the equation

---

[7] The requirement of non-positivity is stronger than absolutely necessary, but is imposed to guarantee that the problem is well posed. Without this, we would need to impose some rather technical growth conditions on u which would be distracting.

[8] Compare (1.20).

$$0 = \tilde{u}(y) - \rho j(y) + (\rho - r)\, y j'(y) + \tfrac{1}{2}|\kappa|^2 y^2 j''(y) \tag{1.48}$$

for j. This is a second-order linear ODE, which is of course much easier to deal with than the non-linear HJB equation which we would have faced if we had tried the value function approach of Section 1.2. We can write the solution of (1.48) as

$$j(y) = j_0(y) + Ay^{-\alpha} + By^{\beta}, \tag{1.49}$$

where $\alpha < 0$ and $\beta > 1$ are the roots of the quadratic

$$Q(t) \equiv \tfrac{1}{2}|\kappa|^2 t(t-1) + (\rho - r)t - \rho, \tag{1.50}$$

and j_0 is a particular solution. How will we find a particular solution? Observe that the equation (1.48) can be expressed as

$$0 = \tilde{u} - (\rho - \mathcal{G})j, \tag{1.51}$$

where $\mathcal{G} \equiv \frac{1}{2}|\kappa|^2 y^2 D^2 + (\rho - r)yD$ is the infinitesimal generator of a log-Brownian motion

$$dY_t = Y_t\{|\kappa|\, dW + (\rho - r)dt\}, \tag{1.52}$$

so one solution would be to take[9]

$$\boxed{j_0(y) = R_\rho \tilde{u}(y) \equiv E\left[\int_0^\infty e^{-\rho t}\tilde{u}(Y_t)\, dt \,\middle|\, Y_0 = y\right],} \tag{1.53}$$

where R_ρ is the ρ-resolvent operator of $\mathcal{G}$. Since $\tilde{u}$ is non-positive decreasing, it is clear that j_0 is also; moreover, since $\tilde{u}$ is convex, and the dynamics for Y are linear, it is easy to see that j_0 must also be convex. The solution j which we seek, of the form (1.49), must be convex, decreasing, and non-positive, so j_0 is a possible candidate, but what can we say about the terms $Ay^{-\alpha} + By^{\beta}$ in (1.49)? By considering the behaviour of j near zero, we see that the only way we can have j (given by (1.49)) staying decreasing and non-positive is if $A = 0$. On the other hand, since j_0 is convex non-positive, it has to be that $|j_0(y)|$ grows at most linearly for large y, and if $B \neq 0$, this would violate[10] either the convexity or the non-positivity of the solution j. We conclude therefore that the only solution of (1.48) which satisfies the required properties is j_0.

Remarks (i) Notice that the special case of CRRA u treated in Section 1.2 works through very simply in this approach; the convex dual $\tilde{u}$ is a power (1.26), and so the expression (1.53) for j_0 can be evaluated explicitly. Verifying that this solution

---

[9] ... provided the integral is finite ...

[10] Recall that $\beta > 1$.

agrees with the earlier solution for this special case is a wholesome exercise; do notice however that the present dual approach is not restricted to CRRA utilities.
(ii) The expression (1.53) for j_0 arose from the probabilistic interpretation of the ODE that j has to solve, but one might suspect that there is a more direct story[11] to explain why this is the correct form. This is indeed the case, and is explained in Section 1.4.

1.4 The Static Programming Approach

We present in this section yet another completely different approach to the basic Merton problem. This one works in greater generality; we do not require the Markovian assumptions of the previous two Sections—the growth rate μ, the volatility σ and the riskless rate r can be any bounded previsible processes.[12] Once again, we suppose that the dividend process δ and the endowment process e are identically zero, but this is for ease of exposition only; it is not hard to extend the story to relax this assumption. We also discuss only the infinite-horizon story for clarity of exposition, but the argument goes through just as well for problems of the form (1.4) as well. The utility[13] u should be strictly concave, strictly increasing in its second argument, continuous in its first, and satisfy (1.39), as in Section 1.3.

There is a price to be paid for the greater level of generality, though; this approach really only works in *complete markets.* In outline, the argument goes as follows. It is not hard to show (without the completeness assumption) that any admissible (θ, c) has to be *budget feasible*, that is, the time-zero value of all future consumption must not exceed the wealth w_0 available at time 0. The key point is that in a complete market, *any budget-feasible consumption stream is admissible* for an appropriate portfolio process θ; this requires the (Brownian) integral representation of a suitable L^1 random variable. Thus we are able to replace the admissibility constraint on (θ, c)—which is after all a dynamic constraint, requiring $w_t \geq 0$ for all t—with a simple static constraint, that the time-zero value of the chosen consumption stream should not exceed the initial wealth.

To begin the detailed argument, we suppose that the asset prices S evolve as

$$dS_t^i = S_t^i\Big(\sum_{j=1}^{N} \sigma_t^{ij} dW_t^j + \mu_t^i dt\Big), \qquad (i = 1, \ldots, N) \tag{1.54}$$

[11] ... one which leads straight to (1.53) without the need to eliminate the spurious solutions to the homogeneous ODE ..

[12] We also require σ^{-1} bounded.

[13] There is nothing to prevent us from having $u(t, x) = u(\omega, t, x)$, where $(\cdot, \cdot, x)$ is a previsible process for each x; the arguments go through without modification.

where σ, σ^{-1}, μ and r are assumed bounded previsible. Next define the *state-price density process* ζ by

$$d\zeta_t = \zeta_t\{-\kappa_t \cdot dW_t - r_t\,dt\}, \qquad \zeta_0 = 1, \tag{1.55}$$

where $\kappa_t \equiv \sigma_t^{-1}(\mu_t - r_t)$ is a bounded previsible process, in view of the assumptions on the coefficient processes. We can express the state-price density process alternatively as[14]

$$\zeta_t = e^{-\int_0^t r_s\,ds} Z_t \equiv e^{-\int_0^t r_s\,ds} \exp\left\{-\int_0^t \kappa_s \cdot dW_s - \tfrac{1}{2}\int_0^t |\kappa_s|^2\,ds\right\}, \tag{1.56}$$

which represents the state-price density as the product of the *discount factor* $\exp(-\int_0^t r_s\,ds)$, which discounts cash values at time t back to time-0 values, and the *change-of-measure martingale*[15] Z. Using this, we can in the usual way define a new probability P^* by the recipe

$$\left.\frac{dP^*}{dP}\right|_{\mathscr{F}_t} = Z_t. \tag{1.57}$$

The change-of-measure martingale Z changes W into a Brownian motion with drift $-\kappa$, which means that the growth rates of the assets in the new measure P^* are all converted to r_t. Thus the discounted asset prices are all martingales in the (pricing) measure P^*. According to arbitrage pricing theory,[16] the time-s price of a contingent claim X_t to be paid at time $t > s$ will be

$$X_s = E^*\big[e^{-\int_s^t r_u\,du} X_t \mid \mathscr{F}_s\big] = \zeta_s^{-1} E\big[\zeta_t X_t \mid \mathscr{F}_s\big]. \tag{1.58}$$

From this, the time-0 price of a consumption stream $(c_t)_{t\geq 0}$ should be calculated as

$$E\left[\int_0^\infty \zeta_s c_s\,ds\right], \tag{1.59}$$

and for this consumption stream to be feasible its time-0 value should not exceed the initial wealth w_0. However, we are able to prove all of this directly, without appeal to general results from arbitrage pricing theory. It goes like this. The process

$$Y_t = \zeta_t w_t + \int_0^t \zeta_s c_s\,ds \tag{1.60}$$

---

[14] Compare this expression with (1.34).

[15] Since κ is bounded, the process Z *is* a martingale, by Novikov's criterion.

[16] Arbitrage pricing theory only requires that discounted assets should be martingales in *some* risk-neutral measure, but under the complete markets assumption, there is only one. We do not actually require anything from arbitrage pricing theory here—everything is derived directly.

is readily verified to be a local martingale, just as we saw at (1.35). Since Y is non-negative, it is a supermartingale, and thus we have the same inequality

$$w_0 = Y_0 \geq E[Y_\infty] \geq E\left[\int_0^\infty \zeta_s c_s \, ds\right]. \tag{1.61}$$

as we had at (1.36). This achieves the first part of the argument, that the time-0 value of an admissible consumption stream cannot exceed the initial wealth.

For the second part, suppose that we are given some non-negative previsible process c which satisfies the budget constraint (1.61). We then define the integrable random variable

$$Y_\infty = \int_0^\infty \zeta_s c_s \, ds$$

and the (uniformly-integrable) martingale

$$Y_t = E[\, Y_\infty \,|\, \mathscr{F}_t \,].$$

By the Brownian martingale representation theorem (see, for example, Theorem IV.36.5 of [34]), for some previsible locally-square-integrable process H we have

$$Y_t = E[Y_\infty] + \int_0^t H_s \, dW_s; \tag{1.62}$$

if we now use the control pair (n, c) defined by

$$n_t S_t = (\sigma_t^T)^{-1}(\zeta_t^{-1} H_t + \kappa_t), \tag{1.63}$$

then the wealth process w generated from initial wealth $w_0' = E[Y_\infty]$ satisfies

$$\zeta_t w_t + \int_0^t \zeta_s c_s \, ds = Y_t = E\left[\int_0^\infty \zeta_s c_s \, ds \,\middle|\, \mathscr{F}_t\right]. \tag{1.64}$$

In particular,

$$\zeta_t w_t = E\left[\int_t^\infty \zeta_s c_s \, ds \,\middle|\, \mathscr{F}_t\right] \geq 0 \tag{1.65}$$

since $c \geq 0$. Thus the control pair (n, c) is admissible.

To summarise, then, *any non-negative consumption stream c satisfying the budget constraint* (1.61) *is admissible; there is a portfolio process n such that the pair* (n, c) *is admissible.*

Using this, the optimization problem

$$\sup_{(n,c)\in\mathscr{A}(w_0)} E\left[\int_0^\infty u(t, c_t) \, dt\right]$$

becomes the optimization problem

$$\sup_{c\geq 0} E\left[\int_0^\infty u(t,c_t)\,dt\right] \quad \text{subject to} \quad E\int_0^\infty \zeta_t c_t\,dt \leq w_0.$$

This is now easy to deal with; absorbing the constraint with a Lagrange multiplier y, we find the problem

$$\sup_{c\geq 0} E\left[\int_0^\infty (u(t,c_t) - y\zeta_t c_t)\,dt\right] + yw_0, \tag{1.66}$$

and we can just optimize this t by t and ω by ω inside the integral: the optimal c^* satisfies

$$u'(t,c_t^*) = y\zeta_t, \tag{1.67}$$

or, equivalently,

$$c_t^* = I(t, y\zeta_t), \tag{1.68}$$

where I is the inverse marginal utility, $I(t,y) = \inf\{x : u'(t,x) < y\}$, a decreasing continuous[17] function of its second argument. This identifies the optimal consumption c, up to knowledge of the multiplier y, which is as usual adjusted to make the contraint hold:

$$E\int_0^\infty \zeta_t I(t, y\zeta_t)\,dt = w_0. \tag{1.69}$$

Because of the assumptions on u, this equation will always have a solution y for any $w_0 > 0$, provided that for some $y > 0$ the left-hand side is finite.

This argument leads us to a candidate for the optimal solution, which needs to be verified of course; but the verification follows exactly as for the verification step in Section 1.2, as the reader is invited to check.

1.5 The Pontryagin-Lagrange Approach

The idea of the Pontryagin-Lagrange approach is to regard the wealth dynamics as a *constraint* to be satisfied by w, n and c. The view taken here is that this should be treated as a principle, not a theorem; while we shall present a plausible argument for why we expect this approach to lead to the solution of the original problem, there are steps on the way that would only hold under technical conditions which would probably be too onerous to check in practice.[18] Our stance is consistent with

[17] ... because of the strict concavity assumption ...

[18] See Rogers [31], Klein & Rogers [22], which apply deep general results of Kramkov & Schachermayer [23] to arrive at a result of this kind.

that announced earlier, namely, we seek solutions, not general results which tell us that solutions are characterized by relations which we cannot solve. In this spirit, the Pontryagin-Lagrange principle is quite capable of leading us to a dual problem, which we may in some situations be able to solve. In any given problem, we would aim to prove optimality by a separate verification argument, rather than trying to tighten up all the steps of the Pontryagin-Lagrange derivation.

To illustrate how this works, we take the standard continuous wealth dynamics

$$dw_t = r_t w_t\, dt + n_t \cdot (dS_t - r_t S_t\, dt) - c_t\, dt \tag{1.70}$$

with no dividend or endowment processes, along with the infinite-horizon objective

$$\sup_{(n,c)\in\mathscr{A}(w_0)} E\left[\int_0^\infty u(t, c_t)\, dt\right]. \tag{1.71}$$

Writing $B_t \equiv \exp(\int_0^t r_s\, ds)$ for the value of the bank account at time t, we have as usual for a self-financing portfolio that

$$w_t = n_t \cdot S_t + \varphi_t B_t, \tag{1.72}$$

where φ_t denotes the number of units of the bank account held at time t (equivalently, the cash value invested in the bank at time t is $\varphi_t B_t$.)

Viewing (1.70) as a constraint on w, c, and n, it is natural to try to absorb this constraint into the objective using the continuous Lagrangian semimartingale λ_t, resulting in the Lagrangian form of the problem:

$$\begin{aligned}
L &\equiv \sup_{n,c,w} E\left[\int_0^\infty \{u(t,c_t)\, dt + \lambda_t\{r_t w_t\, dt + n_t \cdot (dS_t - r_t S_t\, dt) - c_t\, dt\} - \lambda_t dw_t\}\right] \\
&= \sup_{n,w} E\left[\int_0^\infty \{\tilde{u}(t,\lambda_t)\, dt + \lambda_t r_t w_t\, dt + \lambda_t n_t \cdot (dS_t - r_t S_t\, dt) - \lambda_t dw_t\}\right] \\
&= \sup_{n,w} E\left[\int_0^\infty \{\tilde{u}(t,\lambda_t)\, dt + \lambda_t r_t B_t \varphi_t\, dt + \lambda_t n_t \cdot dS_t - \lambda_t dw_t\}\right] \\
&= \sup_{n,w} E\left[\int_0^\infty \{\tilde{u}(t,\lambda_t)\, dt + \lambda_t r_t B_t \varphi_t\, dt + \lambda_t n_t \cdot dS_t + w_t d\lambda_t + dw_t d\lambda_t\} - [\lambda_t w_t]_0^\infty\right] \\
&= \sup_{n,\varphi} E\left[\int_0^\infty \{\tilde{u}(t,\lambda_t)\, dt + n_t \cdot d(\lambda_t S_t) + \varphi_t d(\lambda_t B_t)\} + \lambda_0 w_0\right],
\end{aligned}$$

where we have used optimization[19] over c, (1.72), integration by parts,[20] and collecting terms in the successive steps of the above development. The final expression

---

[19] The optimization over c only yields a finite supremum if $\lambda_t \geq 0$, so this is a dual-feasibility condition on λ. Notice that the optimizing c satisfies $U'(t, c_t) = \lambda_t$.

[20] We assume that $\lim_{t\to\infty} \lambda_t w_t = 0$, a transversality condition.

$$L = \sup_{n,\varphi} E\left[\int_0^\infty \left\{\tilde{u}(t,\lambda_t)\,dt + n_t \cdot d(\lambda_t S_t) + \varphi_t d(\lambda_t B_t)\right\} + \lambda_0 w_0\right]$$

is revealing; since the processes n and φ are unrestricted, we see that the supremum can only be finite if

$$\lambda_t S_t,\ \lambda_t B_t \quad \text{are martingales.} \tag{1.73}$$

This is a *dual-feasibility* condition on the Lagrange multiplier λ, and so by standard Lagrangian duality theory, we expect that the value of the problem will be given by

$$\inf E\left[\int_0^\infty \tilde{u}(t,\lambda_t)\,dt + \lambda_0 w_0\right], \tag{1.74}$$

where the infimum is taken over all non-negative Lagrangian processes satisfying the dual-feasibility condition (1.73). Thus the Pontryagin-Lagrange approach has converted the original primal problem into the dual problem of minimizing the dual objective

$$\Phi(\lambda) \equiv E\left[\int_0^\infty \tilde{u}(t,\lambda_t)\,dt + \lambda_0 w_0\right] \tag{1.75}$$

over dual-feasible λ.

Pause for a moment to think what dual feasibility (1.73) means; if we write $Z_t = \lambda_t B_t$, we have that Z is a positive martingale such that $Z_t S_t / B_t$ is a martingale. Regarding Z as a change-of-measure martingale, changing from measure P to P^*, what this says is that

under P^*, the discounted asset price S/B is a martingale.

Thus P^* is an *equivalent martingale measure*; the dual problem minimizes the dual objective Φ over all equivalent martingale measures.

In general, the class of all equivalent martingale measures may be too large to characterize simply, but one case where it is not is the case of a complete market, where there is just one equivalent martingale measure. This is the situation considered in Section 1.4, where the state-price density process is some multiple of the state-price density ζ defined at (1.55). If we therefore write $\lambda_t = y\zeta_t$ for some $y > 0$, the Pontryagin-Lagrange dual problem is to obtain

$$\inf_{y>0} \Phi(y\zeta) = \inf_{y>0} E\left[\int_0^\infty \tilde{u}(t, y\zeta_t)\,dt + y w_0\right]. \tag{1.76}$$

Elementary calculus gives the derivative of the dual function $\tilde{u}(t,\cdot)$ to be $-I(t,\cdot)$, so if we formally differentiate (1.76), we find that the minimizing y should satisfy

$$E\left[\int_0^\infty \zeta_t I(t, y\zeta_t)\,dt\right] = w_0, \tag{1.77}$$

a relation we saw before at (1.69). During the optimization of the Lagrangian, we found that the optimal consumption c^* is related to the dual variables λ by

$$u'(t, c_t^*) = \lambda_t = y\zeta_t,$$

which is exactly the same as (1.67). It is clear that the static programming approach of Section 1.4 and the current Pontryagin-Lagrange approach are leading to the same place.

1.6 When is the Merton Problem Well Posed?

We return now to the issue raised in Section 1.2 when we calculated the solution to the infinite-horizon Merton problem. We saw that the optimal rule was to invest in the different assets proportionally to wealth, according to the Merton portfolio π_M, and to consume proportionally to wealth, with constant of proportionality

$$\gamma_M = R^{-1}\big\{\rho + (R-1)(r + \tfrac{1}{2}|\kappa|^2/R)\big\}.$$

It is possible that this constant γ_M is negative; *what happens then?* We cannot after all consume at a negative rate in this example! Notice that this can only be an issue if $R \in (0, 1)$; and if $R \in (0, 1)$, the utility is unbounded above. This raises the possibility that the value function of the Merton problem might be infinite, if it were possible to grow the (utility of the) investor's wealth faster than the discounting rate ρ. As we shall demonstrate in the following result, this can indeed happen, and the condition for it *not* to happen is exactly the condition $\gamma_M > 0$.

Proposition 1.3 *The infinite-horizon Merton problem* (1.18) *is well posed*[21] *if and only if* $\gamma_M > 0$.

Proof One implication is already known; if $\gamma_M > 0$, then we derived the optimal solution to the problem in Section 1.2, and by direct calculation (left as an exercise!) were able to find the value of the problem and show that it is finite. For the other implication, we shall suppose firstly that $\gamma_M < 0$, and consider some linear rule for investment and consumption, where we set $\theta_t S_t = \pi w_t$, and $c_t = \gamma w_t$ for some fixed $\pi \in \mathbb{R}^N$ and $\gamma > 0$. Using this policy, the wealth evolves as

$$dw_t = w_t\{r\,dt + \pi \cdot (\sigma dW_t + (\mu - r)\,dt) - \gamma\,dt\},$$

so that

$$w_t = w_0 \exp\{a \cdot W_t + bt\},$$

[21] That is to say, the supremum on the right-hand side of (1.18) defining the value function is finite for any $w_0 > 0$.

where $a = \sigma^T \pi$ and $b = r + \pi \cdot (\mu - r) - \gamma - \frac{1}{2}|a|^2$. Using this rule gives the agent the objective

$$\begin{aligned} \varphi &= u(\gamma w_0)\, E\left[\int_0^\infty \exp(-\rho t + (1-R)a \cdot W_t + (1-R)bt)\, dt\right] \\ &= u(\gamma w_0) \int_0^\infty \exp(-\rho t + (1-R)bt + \tfrac{1}{2}(1-R)^2|a|^2 t)\, dt \end{aligned} \tag{1.78}$$

which will be infinite if

$$\begin{aligned} 0 &\le -\rho + (1-R)b + \tfrac{1}{2}(1-R)^2|a|^2 \\ &= -\left(\rho + (R-1)(b + \tfrac{1}{2}(1-R)|a|^2)\right) \\ &= -\left(\rho + (R-1)(r + \pi \cdot (\mu - r) - \gamma - \tfrac{1}{2}R|a|^2)\right) \end{aligned}$$

If $R \in (0, 1)$, this last expression is maximized over π when $\pi = \pi_M$, to value $-(1-R)\gamma - R\gamma_M$. Since we are assuming that $\gamma_M < 0$, it follows that for small enough $\gamma > 0$ we can generate an infinite value for the objective, and the problem is therefore ill posed.

The final situation is the knife-edge case $\gamma_M = 0$. If we consumed nothing, and invested according to the Merton portfolio, then wealth would evolve as

$$dw_t = w_t\{r\, dt + \pi_M \cdot (\sigma dW_t + (\mu - r)\, dt)\,\};$$

let us write the solution of this SDE as $\bar{w}_t$. From the calculations just done, we know that

$$Ee^{-\rho t} u(\bar{w}_t) = \text{constant} \tag{1.79}$$

for all t. If now consume at time t at rate $\lambda w_t/(1+t)$ for some $\lambda > 0$, we shall have the wealth dynamics

$$dw_t = w_t\{r\, dt + \pi_M \cdot (\sigma dW_t + (\mu - r)\, dt) - \frac{\lambda}{1+t}\, dt\,\};$$

and solution $w_t = \bar{w}_t(1+t)^{-\lambda}$. If we now take $\lambda = (1-R)^{-1}$. Thus the utility of consumption is

$$u(c_t) = u\left(\frac{\lambda w_t}{1+t}\right) = u\left(\frac{\lambda \bar{w}_t}{(1+t)^{1+\lambda}}\right) \propto u(\bar{w}_t)(1+t)^{-(1+\lambda)(1-R)}.$$

If we take $\lambda > 0$ given by

$$1 + \lambda = \frac{1}{1-R},$$

then using the fact (1.79) we learn that

$$E\int_0^\infty e^{-\rho t}u(\bar{w}_t)\,dt \propto \int_0^\infty \frac{dt}{1+t} = \infty,$$

and so the problem is ill posed.

1.7 Linking Optimal Solutions to the State-Price Density

When we solved the infinite-horizon Merton problem in Section 1.2 by the value function approach, we found that the state-price density process ζ, there defined as $\zeta_t = u'(t, c_t^*)$, works as a pricing kernel (see (1.35), (1.37)). Though this appeared from explicit calculation, this property is no accident; indeed, it is a general principle, as we shall now show.[22] We shall discuss the solution to the infinite-horizon problem (1.5) to fix ideas, though the general components work as well for other problems, such as the terminal-wealth problem (1.6).

Suppose that the agent's objective

$$E\left[\int_0^\infty u(t, c_t)\,dt\right]$$

is optimized at $(n^*, c^*) \in \mathscr{A}(w_0)$. Fix $0 \le s < t$, and some event $F \in \mathscr{F}_s$. We consider an infinitesimal change in the agent's policy; if F happens, he decides to consume at rate $c^*_\cdot - \varepsilon S^i_s$ in the interval $(s, s+h)$, where ε and h are small. He uses the money saved to buy himself[23] εh units of asset i, which he then sells at time t for price $\varepsilon h S^i_t$, immediately consuming the proceeds in $(t, t+h)$, raising his consumption rate to $c^*_\cdot + \varepsilon S^i_t$. The overall increase in his objective is to leading order

$$\Delta = E\big[-u'(s, c_s^*)\varepsilon h S^i_s + u'(t, c_t^*)\varepsilon h S^i_t;\ F\big].$$

Since (n^*, c^*) is optimal, this must be zero (to leading order), otherwise it would benefit the agent to do this trade (or its reverse). Since $F \in \mathscr{F}_s$ is arbitrary, we conclude that

$$u'(t, c_s^*)S^i_s = E\big[u'(t, c_t^*)S^i_t \mid \mathscr{F}_s\big]. \tag{1.80}$$

---

[22] The argument that follows would be familiar to any trained economist; see, for example, the account of Breeden [5]. It is a principle, not a theorem, just like the Pontryagin-Lagrange approach; it leads us to useful conclusions and insights, but other methods are needed to confirm these in any given situation.

[23] Of course, it is possible that this perturbed consumption is not feasible; it could become negative at some point in $(s, s+h)$. We ignore such issues, as we may once we let ε, h become infinitesimal.

Thus $\zeta_t \equiv u'(t, c_t^*)$ serves as a state-price density; we have that $\zeta_t S_t^i$ *is a martingale.* Notice that there was nothing special about the ith asset - the argument works for any traded asset. Moreover, there is no need for the market to be complete.

1.8 Dynamic Stochastic General Equilibrium Models

One of the commonest mistakes in mathematical modelling is to model derived quantities, not the fundamentals from which they are derived, and even in mathematical finance, there are plenty of examples of this. The mistake is that because the derived quantities are derived from some common driver, they will in general have to satisfy various inter-relations, and if one simply imposes some arbitrary form for the derived quantities, quite likely the necessary inter-relations will fail. One nice example of this is exposed in Rogers & Tehranchi [35], which shows that[24] it is impossible to have an option implied volatility surface moving by parallel shifts. Another example arises in the HJM approach to interest-rate modelling, which models the instantaneous forward rates. These are derived from a spot rate process, and must satisfy various inter-relations; the drift and volatility of the forward rates cannot be chosen arbitrarily. The original paper of Heath, Jarrow & Morton [18] identifies the appropriate conditions.

At some level, the same applies to the common approach in mathematical finance, which is to begin with the asset price processes. From an economist's point of view, asset prices are *derived* quantities, not fundamentals; *the prices arise as market-clearing prices in a general equilibrium*, and the fundamentals are the fundamentals of the general equilibrium, namely, the agents' preferences, the holdings of the assets, and the output processes of the assets. Typically, general equilibria are hard to solve, and this is one of the reasons that certain features of a general equilibrium[25] may be extracted and used as a starting point for a theoretical discussion. Nonetheless, it is worth attempting to work from fundamentals whenever this is feasible.

Let us see how this works in a dynamic stochastic situation. Suppose[26] that there is just one asset in the world, which delivers an output stream δ_t. There are J agents, whose preferences over consumption streams are given by von Neumann-Morgenstern preferences: agent j orders consumption streams c according to

$$\mathscr{U}_j(c) \equiv E\left[\int_0^\infty u_j(t, c_t)\, dt\right], \tag{1.81}$$

---

[24] ... under minor technical conditions ...

[25] ... for example, supply and demand curves ....

[26] .. for simplicity; the entire analysis goes through with only notational changes if there are multiple assets, and indeed multiple goods.

where as usual $u_j(t,\cdot)$ is increasing, strictly concave, and we shall suppose satisfies the Inada conditions.[27]

We aim to find an equilibrium price process S for the asset, and an equilibrium interest rate r for riskless borrowing. What does this mean? Suppose we are given S and r; then the wealth dynamics faced by each agent are the familiar dynamics of (1.70),

$$dw_t = r_t w_t \, dt + n_t \cdot (dS_t - r_t S_t \, dt) - c_t \, dt.$$

If $(n^j, c^j) \in \mathscr{A}(w_0^j)$ is agent j's choice which maximizes his objective (1.81) given these dynamics, then (S, r) constitute equilbrium prices if *the markets clear:*

$$\sum_j c_t^j = \delta_t, \qquad \sum_j w_t^j = S_t. \tag{1.82}$$

What this means is that at all times, the total output of the asset is being exactly consumed, and the total wealth of all the agents is the asset itself. To amplify, it may be at any time that some of the agents hold some of the asset, and some money in the bank account, but that the aggregate holding of the asset is one unit, and the aggregate holding of money is zero - the net supply of asset is one, the net supply of money is zero.[28] Section 1.7 tells us what the individual agents' optimal solution looks like; we shall have that

$$u_j'(t, c_t^j) = \zeta_t^j \tag{1.83}$$

defines the state-price density for agent j, which determines how agent j prices all assets and contingent claims. Thus agent j will value the productive asset at time t as

$$S_t^j = (\zeta_t^j)^{-1} E\left[\int_t^\infty \zeta_s^j \, \delta_s \, ds \,\middle|\, \mathscr{F}_t \right], \tag{1.84}$$

since holding the asset at time t is equivalent to receiving the output δ_s at all times $s \geq t$. However, since the asset is traded, all agents will agree on its price in equilibrium—otherwise agents would immediately trade the asset on terms which they each considered advantageous.

This is the general solution, and in general it is hard to say much more about it. Even though all agents must agree on the price of the asset, they may disagree on the prices of non-marketed contingent claims; the state-price densities ζ^j are not necessarily the same. If they are not all the same (up to constant multiples), there is not much we can do; if they are, we can make progress. There are in effect three special cases where we can assert that the state-price densities are the same (up to constant multiples):

---

[27] Thus $\lim_{x \downarrow 0} u'(t,x) = \infty$, $\lim_{x \uparrow \infty} u'(t,x) = 0$.

[28] Of course, the net supply of money need not be zero; we just suppose this for now, for the purpose of the discussion.

1. a representative agent equilibrium;
2. a central planner equilibrium;
3. a complete market;

For the first, we assume that $J = 1$; there is just one agent in the economy. In this case, for market clearing we must have[29] that $c_t = \delta_t$, and hence the state-price density is

$$\zeta_t = u'(t, \delta_t); \tag{1.85}$$

see ((1.82), (1.83)). This gives the state-price density explicitly in terms of known quantities, the agent's utility, and the output process. Thus we have an expression (1.84) for the asset price, and depending on the exact form of the problem, it may be possible to evaluate the integrals appearing in (1.84). Likewise, if the market is driven by some Brownian motion, we can identify the equilibrium interest rate process by performing an Itô expansion of ζ_t, since

$$\zeta_t^{-1} d\zeta_t = -r_t dt + d(\text{local martingale}),$$

and hence we can pick out the riskless rate as the finite-variation part of $\zeta^{-1} d\zeta$.

A central planner equilibrium is really a representative agent equilibrium masquerading as something more general. The idea is that a central planner selects as his objective

$$\begin{aligned}\mathscr{U}(c) \equiv &\sup\Big\{\sum_j \alpha_j \mathscr{U}_j(c^j) : \sum_j c_t^j = c_t \;\forall t\Big\}\\ =&E\left[\int_0^\infty \bar{u}(t, c_t)\, dt\right],\end{aligned}$$

where

$$\bar{u}(t, x) \equiv \sup\{\sum_j \alpha_j\, u_j(t, x_j) : \sum_j x_j = x\}, \tag{1.86}$$

and α_j are positive weights. Thus the central planner receives the consumption stream c, and splits it up between the agents so as to maximize the weighted sum of their individual utilities. Solving the optimization problem implicit in the definition of $\bar{u}$ by the Lagrangian approach results in the information that

$$\alpha_j u_j'(t, x_j) = \lambda_t$$

for some positive process λ. Thus if the central planner equilibrium has resulted in assigning consumption stream c^j to agent j, we see from (1.83) that

$$\zeta_t^j = u_j'(t, c_t^j) = \alpha_j^{-1} \lambda_t \tag{1.87}$$

[29] We omit the superfluous index for the sole agent.

and so all the agents have the same state-price density to within a constant multiple. Of course, this means that all the agents value every contingent claim the same. Solving the central planner equilibrium is achieved by solving the representative agent equilibrium with preferences given by $\mathscr{U}$, and then splitting the output between the agents according to the recipe given by the optimization given by (1.86).

The third situation where the equilibrium can be solved explicitly is where there is a complete market, in which case there can be only one state-price density (up to constant mutiples); we have (1.87). Rearranging to make c^j the subject, and then summing over j, we learn that

$$\delta_t = \sum_j c_t^j = \sum_j I_j(t, \eta_j \zeta_t) \tag{1.88}$$

for some positive constants η_j which are to be determined, using the knowledge of the initial wealths w_0^j of the agents. Indeed, we must have that

$$w_0^j = E\left[\int_0^\infty \zeta_t c_t^j \, dt\right] \tag{1.89}$$

$$= E\left[\int_0^\infty \zeta_t I_j(t, \eta_j \zeta_t) \, dt\right], \tag{1.90}$$

and these J conditions determine the J constants η_j, by the Knaster-Kuratowski-Mazurkiewicz theorem (see Border [4]).

Example Suppose that there is just a single asset, whose dividend process is a log-Brownian motion:

$$d\delta_t = \delta_t(\sigma dW_t + \mu dt),$$

a log-Brownian motion with constant drift and constant volatility. We shall write

$$x_t \equiv \log \delta_t = \log \delta_0 + \sigma W_t + (\mu - \tfrac{1}{2}\sigma^2)t.$$

We suppose that the agents' utilities are of the form

$$u_j(t, c) = e^{-\rho t} u_j(c),$$

where ρ is *common* to all, and the u_j are C^2, strictly concave, increasing, and satisfy the Inada conditions. Writing I_j for the inverse of u_j', the market-clearing condition (1.88) gives us the relation

$$\delta_t = \sum_j I_j(\eta_j e^{\rho t} \zeta_t),$$

which can be inverted (only rarely explicitly) to tell us that

$$e^{\rho t}\zeta_t = \tilde{f}(\delta_t) \equiv f(x_t). \tag{1.91}$$

Nevertheless, we can obtain the equilibrium solution remarkably explicitly here; it is straightforward to calculate numerically, for example. We know from (1.84) that the stock price can be expressed as $S_t = \tilde{\varphi}(\delta_t) \equiv \varphi(x_t)$, where

$$\begin{aligned}\tilde{\varphi}(\delta) = \varphi(x) &= \tilde{f}(\delta)^{-1} E\left[\int_0^\infty e^{-\rho t}\delta_t \tilde{f}(\delta_t)\, dt \,\middle|\, \delta_0 = \delta\right] \\ &= f(x)^{-1} E\left[\int_0^\infty e^{-\rho t} e^{x_t} f(x_t)\, dt \,\middle|\, x_0 = x\right]. \end{aligned}\tag{1.92}$$

Since agent j's consumption stream is $c_t^j = I_j(\eta_j e^{\rho t}\zeta_t) \equiv q_j(x_t)$, a function of x_t alone, we can deduce the expression

$$\begin{aligned} w_j(t) = \psi_j(x_t) &= \tilde{f}(\delta_t)^{-1} E\left[\int_t^\infty e^{-\rho(s-t)}\tilde{f}(\delta_s)\, q_j(x_s)\, ds \,\middle|\, \mathscr{F}_t\right] \\ &= f(x_t)^{-1} E\left[\int_t^\infty e^{-\rho(s-t)} f(x_s)\, q_j(x_s)\, ds \,\middle|\, \mathscr{F}_t\right] \end{aligned}\tag{1.93}$$

for the wealth process of agent j. How are we to simplify expressions such as (1.92, 1.93)? The answer is that we can use the *resolvent operator* R_ρ of the diffusion: see [33], III.3. We have then that

$$f(x)\varphi(x) = (R_\rho F)(x) \tag{1.94}$$

$$f(x)\psi_j(x) = (R_\rho Q_j)(x), \tag{1.95}$$

where $F(x) \equiv e^x f(x)$, $Q_j(x) = f(x)q_j(x)$. The resolvent density in this instance is

$$r_\rho(x,y) = r_\rho(0, y-x) = \exp\left(\frac{c(y-x) - |y-x|\sqrt{c^2+2\rho\sigma^2}}{\sigma^2}\right) \Big/ \sqrt{c^2+2\rho\sigma^2}, \tag{1.96}$$

where $c = \mu - \frac{1}{2}\sigma^2$. Hence

$$f(x)\varphi(x) = \int r_\rho(x,y)\, e^y f(y)\, dy = \int r_\rho(0, y-x)\, e^y f(y)\, dy, \tag{1.97}$$

in which we recognize a convolution integral, which can be evaluated numerically using Fast Fourier Transform.

For the special case where all the agents share a common[30] CRRA utility, we have that $I(x) = x^{-1/R}$, and so $f(x) = e^{-Rx}$. In this case we see from (1.92) that $S_t \propto \delta_t$, which gives the *Black-Scholes-Merton model* for a stock paying dividends at a constant proportional rate. Conceptually this is important because it assures us that the most commonly-used model stands on firm economic foundations, even if the representative agent assumption is an over-simplification.

1.9 CRRA Utility and Efficiency

When we come to study various examples in the next chapter, we shall find many situations where the agent's objective is of the infinite-horizon CRRA form (1.18), but the dynamics are somehow altered; for example, the agent may not be allowed to short the stocks; or the interest rate may be random; or the agent may only be able to change his portfolio infrequently. When we solve these problems, we shall typically find a value function $V(w, x)$, where x is some auxiliary variable, which retains the scaling behaviour of the basic Merton problem:

$$V(\lambda w, x) = \lambda^{1-R}\, V(w, x)$$

for all $\lambda > 0$. This means that we may write[31]

$$V(w, x) = V_M(\Theta(x)w)$$

since both sides depend on w as w^{1-R}. The point of doing this is that we may interpret $\Theta(x)$ as the *efficiency* of the modified problem relative to the original Merton problem; an agent with wealth 1 facing the modified problem would achieve the same value as a standard Merton investor with wealth $\Theta(x)$. Often the modified problem will be the Merton problem with additional restrictions or costs, such as transaction costs, or short-selling constraints, and in such situations we shall find (of course!) that $\Theta(x) \in (0, 1)$.

The point of looking at the efficiency of the modified problem is that it standardizes the effect of the modifications in a common fashion across different coefficients of relative risk aversion; simply quoting the value of the modified problem as a number would make it hard to understand just what the effect of the modification had been.

---

[30] This is a representative agent equilibrium; typically these things are easy to solve, but of limited interest because no interaction effects have been modelled.

[31] We use V_M to denote the value function of the Merton problem (1.30).

Chapter 2
Variations

Abstract The second chapter of the book studies a wide range of different examples which are all in some sense variations on the basic Merton examples of Chapter 1. We study what happens when preferences change; or asset dynamics are changed; or objectives are changed.

Throughout this chapter, we shall be looking at variants of the basic Merton problem. Often (but not always) we shall assume that the wealth of the agent evolves as

$$dw_t = rw_t\,dt + \theta_t(\sigma dW_t + (\mu - r)dt) - c_t dt, \tag{2.1}$$

which we shall refer to as the *standard wealth dynamics*. This choice implicitly assumes that there is a single risky asset; multiple risky assets could be handled in most instances with a proof which differs only notationally. Since our interest is always in the new features of the problems being considered, we shall discard the illusory generality afforded by a multi-asset formulation in favour of a more simple notation. However, when there is a significant difference in the multi-asset case, we will take care to distinguish such situations.

Frequently we shall assume that the agent's objective is to obtain

$$\sup_{c,w\geq 0,\theta} E\left[\int_0^\infty e^{-\rho t}u(c_t)\,dt\right], \tag{2.2}$$

which we shall refer to as the *standard objective*.

Where relevant, all of the studies which follow are illustrated by numerical examples. Various parameters have to be set in order to calculate these examples, and unless mention is made to the contrary we shall use the default values

$$\boxed{R = 2, \quad \rho = 0.02, \quad \sigma = 0.35, \quad r = 0.05, \quad \mu = 0.14.} \tag{2.3}$$

L. C. G. Rogers, *Optimal Investment*, SpringerBriefs in Quantitative Finance,
DOI: 10.1007/978-3-642-35202-7_2, © Springer-Verlag Berlin Heidelberg 2013

2.1 The Finite-Horizon Merton Problem

Our first example is a very gentle warm-up exercise. We briefly presented the finite-horizon Merton problem at (1.4), but then proceeded to discuss almost exclusively the more elegant infinite-horizon analogue. But the same techniques work for the finite-horizon problem, and it is useful to record the form that the solution takes. For simplicity, we will suppose that the utility u is separable, and CRRA in consumption. The agent's objective is therefore taken to be

$$\sup E\left[\int_0^T h(t)u(c_t)\,dt + Au(w_T)\right] \tag{2.4}$$

for some strictly positive function h and constant $A > 0$, where $u'(x) = x^{-R}$ for some $R > 0$, $R \neq 1$. Exploiting the scaling properties which are inherited from the CRRA utility, we see that the value function

$$V(t, w) = \sup E\left[\int_t^T h(t)u(c_t)\,dt + Au(w_T)\,\middle|\, w_t = w\right] \tag{2.5}$$

must have the form

$$V(t, w) = f(t)u(w) \tag{2.6}$$

for some function f. The HJB equation for this problem is

$$0 = \sup_{\theta,c}\left[u(t, c) + V_t + V_w(rw + \theta\cdot(\mu - r) - c) + \tfrac{1}{2}\sigma^2\theta^2 V_{ww}\right], \tag{2.7}$$

directly from (1.13). Substituting the scaled form (2.6) into (2.7) gives

$$0 = \sup_{y,q} u(w)\left[\dot{f} + \{r + y(\mu - r) - q\}(1 - R)f - \tfrac{1}{2}R(1 - R)\,\sigma^2 y^2 f + hq^{1-R}\right], \tag{2.8}$$

where we have $y = \theta/w$, $q = c/w$. The optimality conditions are easily seen to be

$$y = \pi_M, \qquad f = hq^{-R},$$

which tells us that

$$\theta_t^* = \pi_M w_t, \qquad c_t^* = w_t\left(\frac{h(t)}{f(t)}\right)^{1/R}; \tag{2.9}$$

investment is exactly as it always has been, but we no longer (in general) consume at a rate which is a constant multiple of wealth.

Substituting these values back into (2.8) gives us a non-linear ODE for the unknown function f:

$$\dot{f} - (R-1)(r + \kappa^2/2R)f + Rf^{1-1/R}h^{1/R} = 0, \qquad f(T) = A. \tag{2.10}$$

If we substitute $f(t) = g(t)^R$, then we get a first-order linear ODE for g which is easily solved to give

$$g(t) = e^{bt}\left[e^{-bT}A^{1/R} + \int_t^T e^{-bs}\,h(s)^{1/R}\,ds\right], \tag{2.11}$$

where $b \equiv (R-1)(r + \kappa^2/2R)/R$.

Remark If we had $h(t) = e^{-\rho t}$, then it is tempting to guess that we should have $f(t) = ae^{-\rho t}$ for some a. However, if we substitute this into (2.10) we find that the ODE is satisfied only if $a = \gamma_M^{-R}$, which would only be correct if $A = e^{-\rho T}\gamma_M^{-R}$. This makes perfect sense; if this happens, then the residual value $Au(w_T)$ is the value of the infinite-horizon problem (see (1.21))!

2.2 Interest-Rate Risk

This time we take the wealth dynamics to be

$$\begin{aligned} dw_t &= r_t w_t dt + \theta(\sigma dW_t + (\mu - r_t)dt) - c_t dt \\ dr_t &= \sigma_r dB_t + \beta(\bar{r} - r_t)dt, \end{aligned}$$

the salient difference being that the riskless rate is no longer supposed constant, but follows a Vasicek process. The parameters σ_r and $\bar{r}$ are constants, and the two Brownian motions W and B are correlated, $dWdB = \eta dt$. The objective will be

$$V(w,r) = \sup E\left[\int_0^\infty e^{-\rho t}u(c_t)\,dt\,\middle|\,w_0 = 0, r_0 = r\right] \tag{2.12}$$

where as usual $u(w) = w^{1-R}/(1-R)$.

A moment's reflection shows that the solution of the Merton problem now will still scale, with the value function taking the form

$$V(w,r) = u(w)f(r).$$

Writing down the HJB equation for this problem, we find (with $c = qw$, $\theta = sw$)

$$\begin{aligned} 0 &= \sup\left[u(c) - \rho V + \tfrac{1}{2}\sigma^2\theta^2 V_{ww} + \sigma\sigma_r\eta\theta V_{wr} + \tfrac{1}{2}\sigma_r^2 V_{rr} + (rw + \theta(\mu - r) - c)V_w + \beta(\bar{r} - r)V_r\right] \\ &= \sup u(w)\left[q^{1-R} - q(1-R)f - \rho f - \tfrac{1}{2}R(1-R)\sigma^2 s^2 f + (1-R)\sigma\sigma_r\eta s f' + \tfrac{1}{2}\sigma_r^2 f'' \right. \\ &\quad \left. + (r + s(\mu - r))(1-R)f + \beta(\bar{r} - r)f'\right]. \end{aligned}$$

Now optimising this over q and s gives us

$$q = f^{-1/R},$$
$$s = \frac{(\mu - r)f + \sigma\sigma_r\eta f'}{\sigma^2 R f}$$

and when substituted back in gives the following second-order ODE for the HJB equations:

$$0 = Rf^{1-1/R} - \rho f + r(1-R)f + (1-R)\frac{\{(\mu - r)f + \sigma\sigma_r\eta f'\}^2}{2\sigma^2 R f} + \tfrac{1}{2}\sigma_r^2 f'' + \beta(\bar{r} - r)f'. \tag{2.13}$$

Numerics. The ODE (2.13) cannot be solved in closed form, but the numerical solution is not particularly difficult. The method used for this example was to use policy improvement (Section 3.6.1), by discretizing the diffusion for r onto an equally-spaced grid centered on r_∞, of width equal to 7 standard deviations[1] of the Vasicek process on either side—so the grid covered the interval $[r_\infty - 7\sigma/\sqrt{2\beta}, r_\infty + 7\sigma/\sqrt{2\beta}]$. The boundary conditions at the two ends were reflecting.

We obtain an interesting plot of efficiency as a function of r: see Fig. 2.1. The parameter values are $\sigma_r = 0.01$, $\bar{r} = 0.04828$, $\beta = 0.2$ and $\eta = 0.45$, with other parameters taking the default values (2.3). The surprising thing is that the efficiency[2]

Fig. 2.1 Efficiency with Vasicek short rate model

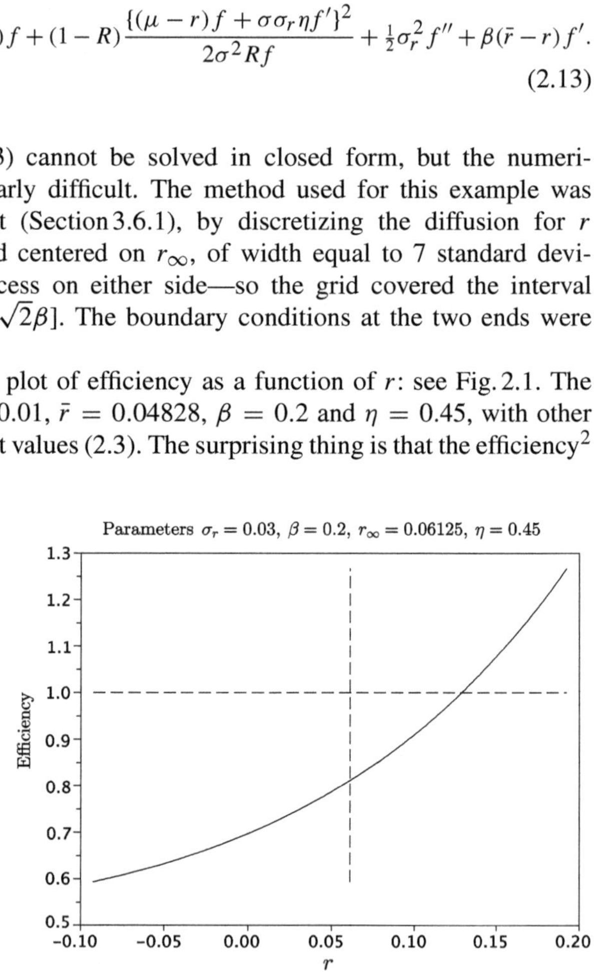

[1] As a check of the effect of the assumed boundary conditions, I calculated the efficiency at $r = 0$, which came out to 0.6972201 using a 7-standard deviation grid, and a 9-standard deviation grid, and a 5-standard deviation grid.

[2] Compared with the Merton problem where we assume that $r = \bar{r}$.

is *greater than 1*! But a moment's thought shows that this may indeed be expected. Part of the effect of the variable interest rate is to make the excess rate of return $\mu - r$ stochastic. Now the dependence of the Merton value on the excess rate of return is *convex*,[3] and so we should expect that the value of the averaged Merton problem will be better than the value for the Merton problem with the average value for excess return—Jensen! Of course, there are differences also in the effects of discounting, so this argument is not conclusive, but it does at least indicate a mechanism which could account for efficiencies in excess of 1. Another possible mechanism would be that if r was very high, then the agent could earn a lot from riskless investment at least for a while before the interest rate reverted back to its long-run average level.

2.3 A Habit Formation Model

Constantinides [8] proposed a model where the agent's consumption is compared to an exponentially-weighted historical average of past consumption. One motivation for this was to try to explain the equity premium puzzle (EPP). The model proposed by Constantinides helps a bit in explaining the EPP, but it is in any case an interesting attempt to explore different objectives. The dynamics taken are a simple variant of the usual wealth equation:

$$dw_t = rw_t dt + \theta_t(\sigma dW_t + (\mu - r)dt) - c_t dt \tag{2.14}$$

$$d\bar{c}_t = \lambda(c_t - \bar{c}_t)dt. \tag{2.15}$$

The agent's objective in Constantinides' account is

$$\sup E \int_0^\infty e^{-\rho t} u(c_t - \bar{c}_t)\, dt$$

so that present consumption is in some sense evaluated relative to the exponentially-weighted (EW) average $\bar{c}_t$ of past consumption. If we use a CRRA utility u, then what we find is that the consumption may never fall below $\bar{c}$, so the agent must keep $\bar{c}_t/r$ in the bank account to guarantee that level of consumption, and then he invests the remaining wealth $w_t - \bar{c}_t/r$. very much as before; the equations are very easy to derive, and we leave them to the reader as an exercise.

What we propose to do here is to keep the dynamics (2.14) and (2.15), but to take as the objective

$$V(w, \bar{c}) \equiv \sup E\left[\int_0^\infty e^{-\rho t} u(c_t/\bar{c}_t)\, dt \,\middle|\, w_0 = w, \bar{c}_0 = \bar{c}\right] \tag{2.16}$$

[3] ... at least in the case $R > 1$ which we deal with here.

which (more realistically) rewards the *ratio* of current consumption to the EW average. This objective permits current consumption to fall below the EW average of past consumption at various times, again a more realistic feature.

The problem does not now admit a simple closed-form solution, in contrast to the problem studied by Constantinides, but there is an obvious scaling, for any $\alpha > 0$:

$$V(\alpha w, \alpha \bar{c}) = V(w, \bar{c}),$$

which allows us to write more simply

$$V(w, \bar{c}) = V(w/\bar{c}, 1) \equiv v(w/\bar{c}). \tag{2.17}$$

The solution is a function of the scaled variable $x_t \equiv w_t/\bar{c}_t$ alone, so we must first understand how this process evolves. We introduce the notation $q_t = c_t/\bar{c}_t$ for the scaled consumption rate. Some routine calculations with Itô's formula give us the dynamics of x:

$$dx_t = rx_t\, dt + \varphi_t(\sigma dW_t + (\mu - r)dt) - (\lambda x_t + 1)q_t dt + \lambda x_t dt, \tag{2.18}$$

where $\varphi = \theta/\bar{c}$. This dynamic is interesting because, although the dependence on the portfolio variable φ is conventional, the dependence on the consumption variable q is not. One observation should be made straight away. It is always a feasible strategy to come out of the risky asset completely ($\varphi \equiv 0$), and to maintain x at its current level; from (2.18), this implies that we could maintain q at the constant value

$$q^{(0)} = \frac{(\lambda + r)x}{1 + \lambda x} \tag{2.19}$$

forever, guaranteeing that the value of the problem would be $\rho^{-1}u(q^{(0)})$. So the value is bounded below by

$$v(x) \geq \rho^{-1}\, u\left(\frac{(\lambda + r)x}{1 + \lambda x}\right). \tag{2.20}$$

For very small x, we would expect that the portfolio φ would have to be small, since x has to be kept non-negative, and if φ remained bounded away from zero as $x \downarrow 0$, the volatility arising from the investment in the risky asset would carry x below zero. This gives us the boundary condition

$$\lim_{x \downarrow 0} v(x)/u(x) = \rho^{-1}\, (\lambda + r)^{1-R}. \tag{2.21}$$

We have reduced the problem to finding

$$v(x) \equiv \sup_{\varphi, q} E\left[\int_0^\infty e^{-\rho t} u(q_t)\, dt \,\middle|\, x_0 = x\right] \tag{2.22}$$

where x evolves as (2.18). In these terms, the HJB equations become more simply

$$\sup_{\varphi,q}\left[u(q) - \rho v + \{rx + \varphi(\mu - r) - (1+\lambda x)q + \lambda x\}\, v' + \tfrac{1}{2}\varphi^2\sigma^2 v''\right] = 0. \quad (2.23)$$

As usual, optimal values of q and φ are found explicitly from

$$u'(q) = (1+\lambda x)v'(x), \quad \varphi = -\kappa v'/\sigma v''.$$

We can transform to the dual equation (via $z \equiv v'(x)$, $J(z) = v(x) - xz$), but the second-order ODE which results:

$$\tilde{u}(z(1-\lambda J')) - \rho J + (\rho - r - \lambda)zJ' + \tfrac{1}{2}\kappa^2 z^2 J'' = 0 \qquad (2.24)$$

no longer admits a closed-form solution, so we are forced down a numerical path.

Numerics. Two different numerical solution methods were used here, and their results compared. The first was policy improvement, where we insisted that the lower bound (2.20) holds with equality at the two ends of the grid. The policy improvement algorithm is therefore solving a Markov decision process which gets stopped at the ends of the interval. The second numerical scheme was to solve the dual HJB equation (2.24) by introducing the variable $s \equiv \log z$, which transforms the linear differential operator into a constant-coefficient form. Then the Newton method (see Section 3.6.3) was applied to calculate the solution, with natural boundary conditions at the two ends of the interval. The value of λ used was $\lambda = 1$. As a diagnostic for comparison, we calculated the numerical value of θ when $x = 1$; the two methods agreed in the first five significant figures. The results are plotted in Fig. 2.2.

The plots reveal very plausible behaviour, which accords more with the behaviour we would expect than the predictions of the basic Merton model. As wealth w rises, we see that the level of current consumption rises quite rapidly to begin with, but then levels off. It requires a lot of wealth before the agent is ready to consume above the averaged value $\bar{c}$, as would be expected; increasing consumption has impact on the future in that we will want to consume more in future to stay as happy, so we are cautious about taking on that additional consumption. In fact, if we wanted to maintain $c/\bar{c}$ constant at some level q, we see from (2.15) that c_t would have to grow as $\exp(\lambda(q-1)t)$, an *exponential* growth of consumption. If q were so large that $\lambda(q-1) > r$ then no initial wealth would be sufficient to support such consumption. While this is not a conclusive analysis, it strongly suggests that the value is bounded above by some *strictly negative*[4] constant. The plot of $\log((1-R)v)$ against $\log(x)$ fits with this; for small values of x what we see looks like a power law, but for large values of x we appear to have convergence to a lower bound.

We also see that as wealth rises our consumption as a fraction of current wealth falls, dropping to limit 0, again entirely as we would expect. As wealth rises, we see

---

[4] Recall (2.3) that we are using $R = 2$.

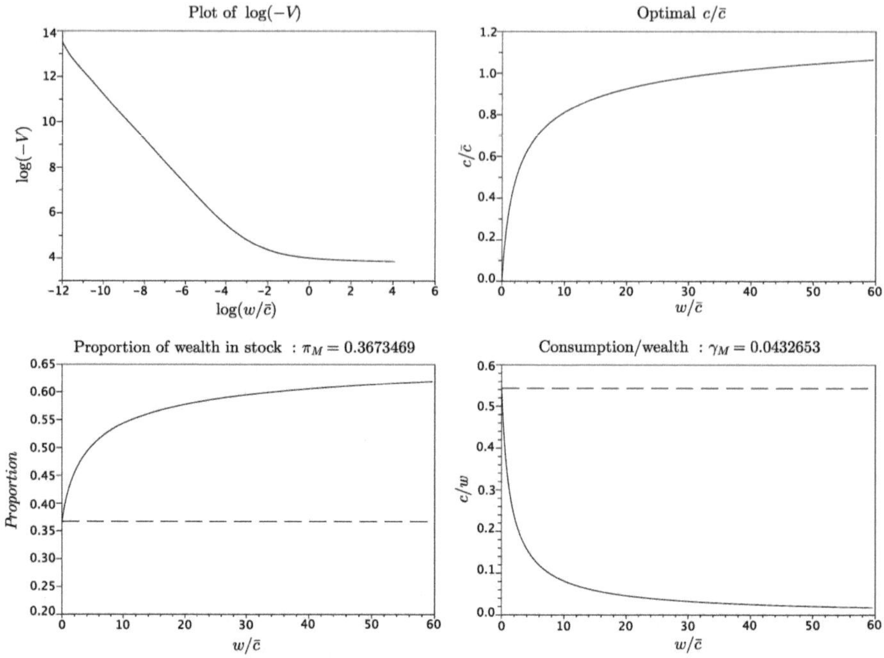

Fig. 2.2 Solution of the habit formation problem, $\lambda = 1$

that the fraction of wealth invested in the stock also goes up; plots calculated over a larger interval show θ/w levelling off at just below 70 %. This is again what we would expect; a wealthy individual can be quite relaxed about risk and would be prepared to venture more in risky ventures. We see that the proportion invested in the risky asset is always higher than the Merton proportion. Similarly, the ratio $c/\bar{c}$ rises gradually with wealth, levelling off at around the level 1.4.

2.4 Transaction Costs

Consider the situation where

$$dX_t = rX_t dt + (1-\varepsilon)dM_t - (1+\varepsilon)dL_t - c_t dt$$
$$dY_t = Y_t(\sigma dW_t + \mu dt) - dM_t + dL_t,$$

where X_t is value of holding of cash, Y_t is value of holding of stock at time t, M_t (respectively, L_t) the cumulative sales (respectively, purchases) of stock by time t. The investor's goal is to achieve

$$V(x,y) = \sup E\left[\int_0^\infty e^{-\rho t} u(c_t)dt \,\middle|\, X_0 = x, Y_0 = y\right],$$

with $u(x) = x^{1-R}/(1-R)$ as in the Merton problem. Using the MPOC, we develop the (super)martingale

$$Z_t = e^{-\rho t} V(X_t, Y_t) + \int_0^t e^{-\rho t} u(c_t)\, dt \tag{2.25}$$

using Itô's formula to learn that

$$\begin{aligned} e^{\rho t}\, dZ_t &\doteq \big\{-\rho V + (rx - c)V_x + \mu y V_y + \tfrac{1}{2}\sigma^2 y^2 V_{yy} - u(c)\big\}\, dt \\ &\quad + \big[V_y - (1+\varepsilon)V_x\big]\, dL + \big[(1-\varepsilon)V_x - V_y\big]\, dM, \end{aligned} \tag{2.26}$$

where $\doteq$ signifies that the two sides differ by a (local) martingale. Since Z must be a supermartingale always, and a martingale under optimal control, we deduce that the three drift terms must be non-increasing. Therefore the HJB equations here are three equations,

$$\begin{aligned} &\sup\Big[u(c) - \rho V + \tfrac{1}{2}\sigma^2 y^2 V_{yy} + \mu y V_y + (rx - c)V_x\Big] \le 0, \\ &(1-\varepsilon)V_x \le V_y \le (1+\varepsilon)V_x. \end{aligned}$$

We shall once again have scaling, so if we set $V(x, y) = y^{1-R} f(p)$, where $p \equiv x/y$, we can re-express this as

$$\begin{aligned} 0 &= \tilde{u}(f') + \tfrac{1}{2}\sigma^2 p^2 f''(p) + (\sigma^2 R - \mu + r)pf'(p) \\ &\quad + \{\mu(1-R) - \rho - \tfrac{1}{2}\sigma^2 R(1-R)\} f(p), \\ &(1-\varepsilon)f' \le (1-R)f - pf'(p) \le (1+\varepsilon)f'. \end{aligned}$$

Alternatively, if we write $f(p) \equiv g(\log(p))$, we simplify the HJB differential operator quite a bit:

$$0 \ge e^{-t(1-1/R)}\tilde{u}(g'(t)) + a_2 g''(t) + a_1 g'(t) + a_0 g(t) - \rho g(t), \tag{2.27}$$

$$0 \ge (1 - \varepsilon + e^t)g'(t) - (1-R)e^t g(t), \tag{2.28}$$

$$0 \ge -(1 + \varepsilon + e^t)g'(t) + (1-R)e^t g(t), \tag{2.29}$$

where $t \equiv \log(p)$, and

$$\begin{aligned} a_2 &= \tfrac{1}{2}\sigma^2, \\ a_1 &= (\sigma^2 R + r - \mu - \tfrac{1}{2}\sigma^2), \\ a_0 &= (R-1)(\tfrac{1}{2}\sigma^2 R - \mu). \end{aligned}$$

Constantinides [7] solves a simplified form of this problem, and Davis & Norman [10] analyse it quite completely. The main conclusion is that there is some *no-trade interval* $K = [t_s, t_b]$ for t such that while t remains within $[t_s, t_b]$, you make no change in your portfolio; if ever $t < t_s$ you immediately sell enough stock to move back into the interval K, and if ever $t > t_b$ you immediately buy sufficient stock to move t back into the interval K.

No closed-form solution is known, but Davis & Norman show how the ODE for g may be solved by iteratively solving the ODE with different initial conditions until the solution closes in on one which satisfies the C^2 pasting condition at the ends of K. The solution method used here is policy improvement. In more detail, suppose that we currently have a policy that we shall buy stock when $t \in \Omega_b$, sell stock when $t \in \Omega_s$, and elsewhere we shall consume at rate $c = yh(t)$. We then find that we have to solve the (linear) ODE

$$\begin{aligned} 0 = \Phi_s(g,t) &\equiv \quad (1-\varepsilon+e^t)g'(t) - (1-R)e^t g(t) \quad (t \in \Omega_s), \\ 0 = \Phi_b(g,t) &\equiv -(1+\varepsilon+e^t)g'(t) + (1-R)e^t g(t) \quad (t \in \Omega_b), \\ 0 = \Phi_0(g,t;h) &\equiv \left\{u(h(t)) - h(t)e^{-t}g'(t)\right\} + a_2 g''(t) + a_1 g'(t) + a_0 g(t) - \rho g(t) \qquad \text{else.} \end{aligned}$$

Having solved this for g, we then go back and compute the functionals $\Phi_s(g,t)$, $\Phi_b(g,t)$, $\sup_h \Phi_0(g,t;h)$ for *all* t, and update the policy according to what we find; we choose to sell in the region where $\Phi_s(g,t)$ is the largest, buy in the region where $\Phi_b(g,t)$ is largest, and elsewhere we consume at rate given by the maximising value of h.

We show in Figs. 2.3 and 2.4 what the solution looks like for this problem, for the default values (2.3) taking $\varepsilon = 0.005$. The first plot, Fig. 2.3 displays the three inequalities (2.27), (2.28) and (2.29) at once the solution has been found, and Fig. 2.4 shows the form of g found, with the changeover points shown by the vertical broken lines. The Merton proportion for this problem is 36.73 %, and we sell stock when the fraction of our wealth in stock rises to 41.98 %, we buy when it falls to 30.39 %. This no-trade interval is remarkably wide, bearing in mind that the proportional transaction cost was only 0.5 %. In fact, the loss of efficiency is $O(\varepsilon^{\frac{2}{3}})$—see [32, 38]. This last tells us that when we consider typical values for the transaction

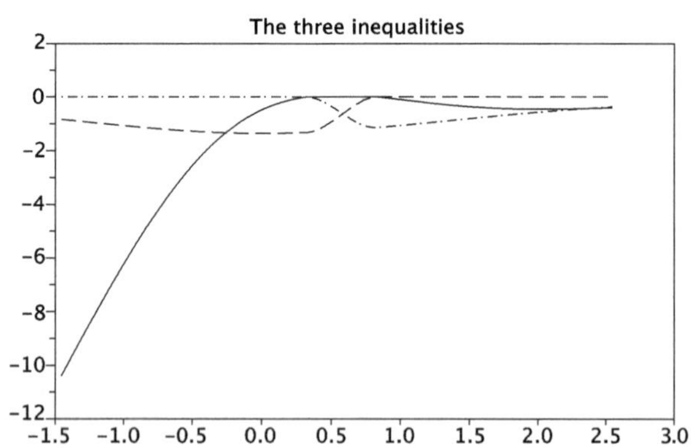

Fig. 2.3 The three inequalities for the transaction costs example

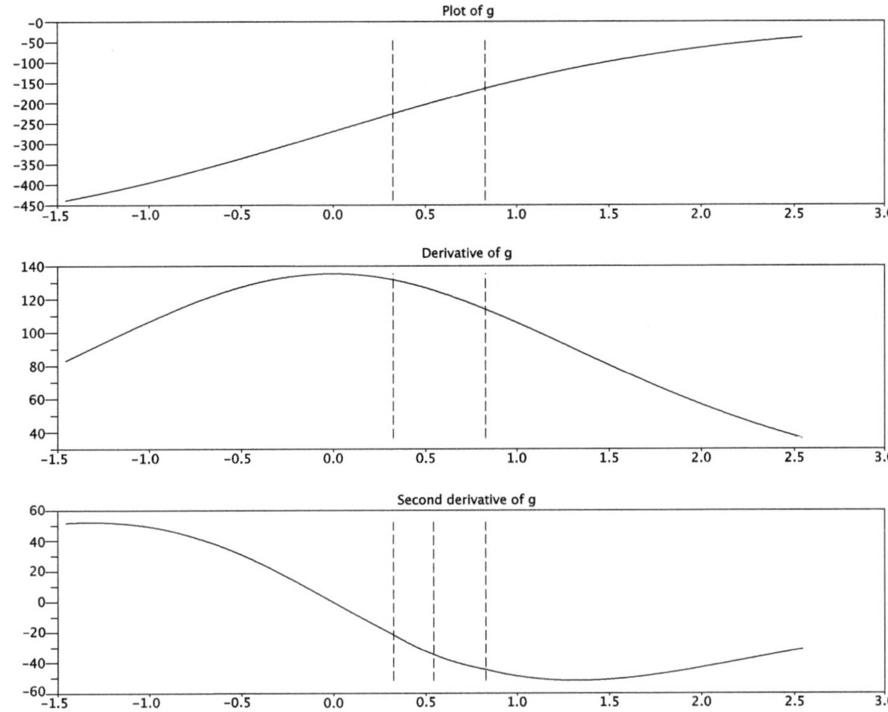

Fig. 2.4 The function g and its first two derivatives

cost (of the order of 1% or less), the impact on efficiency will be *small*, even though the optimal trading policy will look very different from the Merton rule.

2.5 Optimisation under Drawdown Constraints

In this problem, which you will find treated thoroughly by Elie & Touzi [13], we assume the (by now) standard dynamics

$$dw_t = r(w_t - \theta_t)dt + \theta_t(\sigma dW_t + \mu dt) - c_t dt$$

for the wealth and objective

$$\sup E\Big[\int_0^\infty e^{-\rho t} u(c_t) dt\Big], \quad u'(x) = x^{-R},$$

but now we shall impose the constraint

$$w_t \geq b\bar{w}_t = b \sup_{s \leq t} w_s, \quad \forall t, \tag{2.30}$$

where $b \in (0, 1)$ is fixed. This is called a *drawdown constraint*, in a natural terminology. Drawdown constraints are of practical importance for fund managers, because if their portfolio loses too much of its value, the investors are likely to take their money out and that is the end of the story, however clever (or even optimal!) the rule being used by the fund manager. For this problem, the value function

$$V(w, \bar{w}) = \sup E\Big[\int_0^\infty e^{-\rho t} u(c_t)\, dt \,\Big|\, w_0 = w, \bar{w}_0 = \bar{w}\Big]$$

evidently scales like

$$V(w, \bar{w}) = \bar{w}^{1-R}\, V(w/\bar{w}, 1) = \bar{w}^{1-R}\, v(w/\bar{w}) = \bar{w}^{1-R}\, v(x), \quad x = w/\bar{w} \in [b, 1].$$

So the HJB equation here is

$$\sup_{c,\theta} \Big[u(c) - \rho V + \frac{1}{2}\sigma^2\theta^2 V_{ww} + (r(w-\theta) + \mu\theta - c)V_w \Big] = 0$$

with the boundary condition that $V_{\bar{w}} = 0$ at $w = \bar{w}$. Thus the HJB equation is

$$\tilde{u}(V_w) - \rho V + rwV_w - \frac{1}{2}\kappa^2 \frac{V_w^2}{V_{ww}} = 0,$$

where as before $\kappa = (\mu - r)/\sigma$. In terms of v this gives

$$\tilde{u}(v') - \rho v + rxv' - \frac{1}{2}\kappa^2 \frac{(v')^2}{v''} = 0, \tag{2.31}$$

$$(1-R)v(1) = v'(1) \tag{2.32}$$

(indeed, $(1-R)v(x) - xv'(x) \leq 0$ always, with equality when $x \geq 1$). The boundary condition at 1 can be understood as saying that we extend v to $(1, \infty)$ by $v(x) = x^{1-R}v(1)$ $(x \geq 1)$, and this extension is C^1.

The solution of this problem is achieved by using the dual variable technique of Section 1.3: setting

$$z \equiv v'(w)$$

as the new variable, and

$$J(z) = v(w) - wz$$

as the new function, then as a little calculus confirms, we have

$$J'(z) = -w, \quad J''(z) = -1/v''(w).$$

Now (2.31) becomes simply

$$\tilde{u}(z) + \tfrac{1}{2}\kappa^2 z^2 J'' + (\rho - r) z J' - \rho J = 0, \tag{2.33}$$

$$-\left(1 - \frac{1}{R}\right) J(z) + z J'(z) \leq 0, \tag{2.34}$$

with equality in (2.34) when $J'(z) \leq -1$.

One other observation is required: as $w \downarrow b\bar{w}$, the portfolio weight $\theta \to 0$, because otherwise at the boundary the constraint (2.30) would get violated. But recall that the optimal portfolio is

$$\theta = \frac{(\mu - r) V_w}{\sigma^2 V_{ww}};$$

this implies that $v''(b) = +\infty$, $J''(v'(b)) = 0$. Thus there exist $z_b = v'(b) > z_1 = v'(1)$ such that the solution J has the form

$$J(z) = \begin{cases} A_0 \tilde{u}(z) & \text{for } z \leq z_1; \\ A_1 (z/z_b)^{-\alpha} + B_1 (z/z_b)^{\beta} + q\tilde{u}(z) & \text{for } z_1 \leq z \leq z_b; \\ q\tilde{u}(z_b) + A_1 + B_1 + b(z_b - z) & \text{for } z \geq z_b \end{cases}$$

where $q = -1/Q(1 - R^{-1})$, and $Q(t) \equiv \frac{1}{2}\kappa^2 t(t-1) + (\rho - r)t - \rho$ is the quadratic whose roots are $-\alpha < 0 < \beta$. In order that the problem is well posed, it is necessary and sufficient that $q > 0$. The constants A_0, A_1, B_1, z_1, and z_b are to be determined from the conditions

(i) J is C^2 at z_b;
(ii) J is C^1 at z_1.

Thus if we pick z_b, we know that $J'(z_b) = -b$, $J''(z_b) = 0$, so the ODE (2.33) gives us

$$\rho J(z_b) = -(\rho - r) z_b b + \tilde{u}(z_b).$$

We also have the condition that $J'(z_1) = -1 = -A_0 z_1^{-1/R}$, giving us the relation $z_1 = A_0^R$. Using these conditions it is not too hard to find (numerically) the solution J, and hence the original value function v.

To explain in a little more detail, the ratio $J(z)/\tilde{u}(z)$ must be constant in $(0, z_1)$ and is C^1 at z_1. Examining the derivative of this ratio at z_1 gives us the equation

$$\frac{\alpha + 1 - 1/R}{\beta - 1 + 1/R} = \left(\frac{z_1}{z_b}\right)^{\alpha+\beta}.$$

We are therefore able to deduce the value of z_1 given z_b. But since $A_0 = z_1^{1/R}$, we now know what the value of J must be on the left at z_1, and we adjust the value of z_b until we have continuity at z_1.

We see in Fig. 2.5 the solution when taking $b = 0.8$, other parameters as at (2.3). The efficiency in this case has fallen to 90.06 %, representing a fairly substantial loss. If we chose $b = 0.6$, for example, the efficiency would be 96.53 %. The impact on investment and consumption is also very noticeable. The Merton proportion is 36.73 %, but under the drawdown constraint the fraction of wealth in the risky asset never exceeds 27 %. The optimal rate for consuming in the Merton problem is 4.326 %, but with the drawdown constraint it reaches a maximal value of 4.106 % only. While it may feel like a good idea to insist on a drawdown constraint, not many funds would operate drawdown control in the way this example recommends; doing so has an unpleasant tendency to lock in losses.

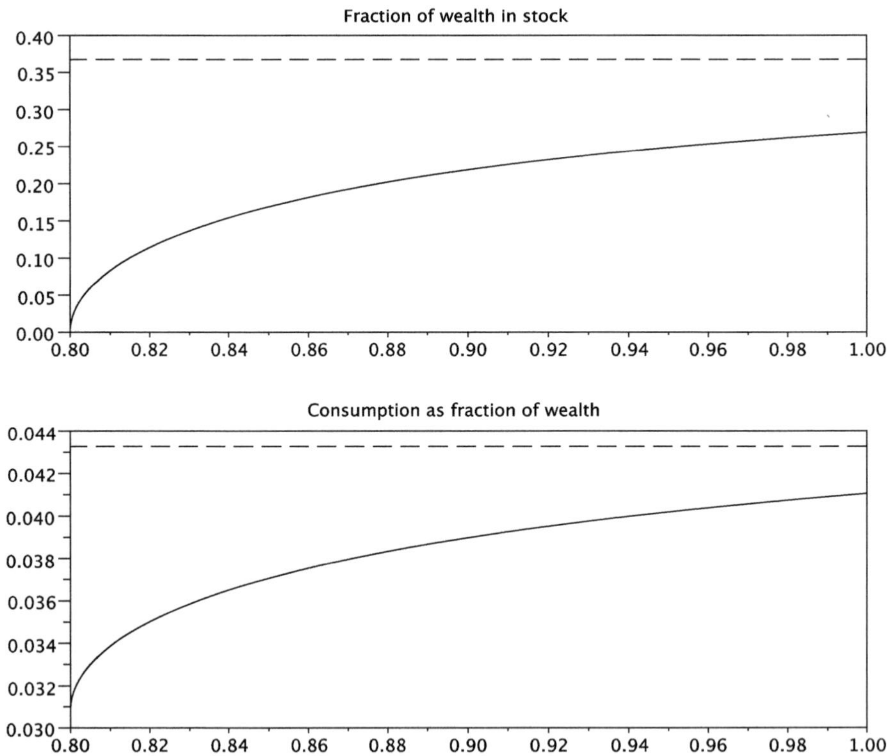

Fig. 2.5 Investment in stock and consumption rate as a function of $w/\bar{w}$

2.6 Annual Tax Accounting

What is the effect on the Merton problem of an annual tax on capital gains? Suppose that u is again CRRA, and at each time $t = nh$ we have to pay tax on wealth gain over the last time period of length h. Thus $w_{nh} = w_{nh-} - \tau(w_{nh-} - w_{nh-h}) = (1-\tau)w_{nh-} + \tau w_{nh-h}$. If we do this, then the problem becomes a finite-horizon problem,

$$V(w) = \sup E\left[\int_0^h e^{-\rho s}u(c_s)ds + e^{-\rho h}u(\tau w + (1-\tau)w_h)\right].$$

Clearly by scaling again, there is some positive constant A such that $V(w) = Au(w)$, so we have to consider

$$\sup E\left[\int_0^h e^{-\rho s}u(c_s)ds + Ae^{-\rho h}u(\tau w + (1-\tau)w_h)\right].$$

As we saw in Section 1.4, by (1.67) the optimal terminal wealth w_h^* and running consumption c^* are related to the state-price density process ζ by

$$e^{-\rho t}u'(c_t^*) = e^{-rt}Z_t = \lambda\zeta_t, \quad Z_t = E_t[e^{rh}Ae^{-\rho h}(1-\tau)u'(\tau w + (1-\tau)w_h^*)],$$

where $\zeta_t = \exp\{-rt - \kappa W_t - \frac{1}{2}\kappa^2 t\}$ is the state-price density, $\zeta_0 = 1$. We deduce that $c_t^* = I(\lambda e^{\rho t}\zeta_t)$ and

$$\lambda\zeta_h = e^{-rh}Z_h = Ae^{-\rho h}(1-\tau)u'(\tau w + (1-\tau)w_h^*);$$

rearranging to make w_h^* the subject of the equation gives us

$$w_h^* = \frac{1}{1-\tau}\left\{-\tau w + I\left(\frac{\lambda e^{\rho h}\zeta_h}{A(1-\tau)}\right)\right\}.$$

We now need to relate λ to initial wealth w:

$$\begin{aligned} w &= E\left[\int_0^h \zeta_u c_u^* du + \zeta_h w_h^*\right] \\ &= E\left[\int_0^h \zeta_t^{1-1/R}\lambda^{-1/R}e^{-\rho t/R}dt - \zeta_h\frac{\tau w}{1-\tau} + \frac{\zeta_h^{1-1/R}}{1-\tau}\lambda^{-1/R}e^{-\rho h/R}A^{1/R}(1-\tau)^{1/R}\right] \\ &= -\frac{\tau w e^{-rh}}{1-r} + \lambda^{-1/R}\frac{1-e^{-\gamma h}}{\gamma} + \lambda^{-1/R}A^{1/R}(1-\tau)^{1/R-1}e^{-\gamma h}. \end{aligned}$$

Thus

$$w\left(1+\frac{\tau e^{-\tau h}}{1-\tau}\right)=\lambda^{-1/R}\left(\frac{1-e^{-\gamma h}}{\gamma}+A^{1/R}(1-\tau)^{1/R-1}e^{-\gamma h}\right). \tag{2.35}$$

Now we need to compute the value,

$$\begin{aligned}
V(w) &= E\left[\int_0^h e^{-\rho t}u(c_t^*)dt + Ae^{-\rho h}u(\tau w+(1-\tau)w_h^*)\right]\\
&= E\left[\int_0^h e^{-\rho t}\frac{(\lambda e^{\rho t}\zeta_t)^{1-1/R}}{1-R}dt + \frac{Ae^{-\rho h}}{1-R}\left(\frac{\lambda e^{\rho h}\zeta_h}{A(1-\tau)}\right)^{1-1/R}\right]\\
&= \frac{\lambda^{1-1/R}}{1-R}E\left[\int_0^h e^{-\rho t/R}\zeta_t^{1-1/R}dt + A^{1/R}e^{-\rho h/R}(1-\tau)^{1/R-1}\zeta_h^{1-1/R}\right]\\
&= \frac{\lambda^{1-1/R}}{1-R}\left(\frac{1-e^{-\gamma h}}{\gamma}+A^{1/R}(1-\tau)^{1/R-1}e^{-\gamma h}\right).
\end{aligned} \tag{2.36}$$

Now from the Eq. (2.35), $\lambda^{-1/R} = Bw/K$, where $B = 1+\tau e^{-rh}/(1-\tau)$ and $K = \gamma^{-1}(1-e^{-\gamma h}) + A^{1/R}(1-\tau)^{1/R-1}e^{-\gamma h}$, so we have that $\lambda = (Bw/K)^{-R}$, and from (2.36) we deduce that

$$V(w) = u(w)\left(\frac{B}{K}\right)^{1-R}K = u(w)B^{1-R}K^R = Au(w).$$

This implies that

$$A^{1/R} = KB^{1/R-1} = B^{(1-R)/R}\left(\frac{1-e^{-\gamma h}}{\gamma}+A^{1/R}(1-\tau)^{\frac{1-R}{R}}e^{-\gamma h}\right).$$

We can now make $A^{1/R}$ the subject of this equation:

$$A^{1/R} = \frac{\gamma^{-1}(1-e^{-\gamma h})B^{1/R-1}}{1-e^{-\gamma h}(((1-\tau)B)^{1/R-1})},$$

expressing A (and hence the value) explicitly in terms of the variables of the problem. The efficiency can now be expressed explicitly as

$$\theta = (A\gamma^R)^{1/(1-R)}.$$

We are now able compute numerical values quite explicitly. Figure 2.6 exhibits the remarkable conclusion that for higher tax rates, the efficiency can actually be *greater* than 1! Though it appears counterintuitive, it is not wrong. The effect of tax is to reduce the mean of the net gain by time h, but it also reduces the variance of the net gain. Smaller mean is bad, but smaller variance is good, and these two effects

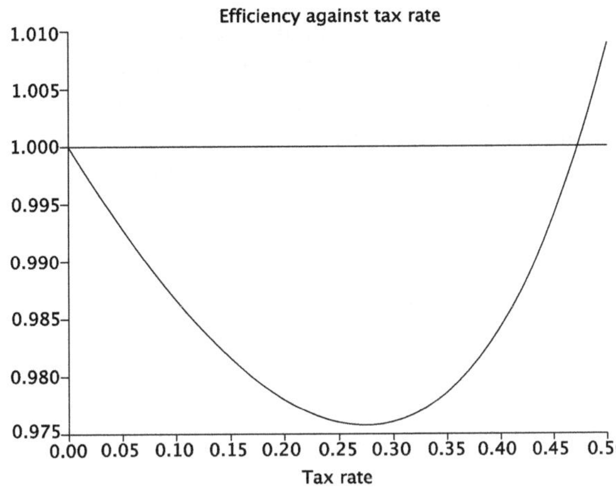

Fig. 2.6 Efficiency as it depends on the tax rate

act against each other. Eventually the improvement due to smaller variance prevails, and the efficiency begins to rise again as tax increases. Notice that the changes in efficiency are in any case quite small. Notice also that the story we have told here is unrealistic; an investor does not get a tax repayment if he makes a loss, he gets a tax credit which he can carry forward to offset against tax he would have to pay on future profits. This makes the story more complicated.

2.7 History-Dependent Preferences

This is an attempt to make a model where preferences depend somehow on integrated consumption over a period, rather than just a consumption rate. We postulate the dynamics

$$dw_t = rw_t dt + \theta_t(\sigma dW_t + (\mu - r)dt) - c_t dt \tag{2.37}$$

$$d\xi_t = \lambda(c_t^\alpha - \xi_t)dt. \tag{2.38}$$

Here, $\lambda > 0$ and $\alpha \in (0, 1)$ are constants. The process ξ has the representation

$$\xi_t = \int_{-\infty}^{t} \lambda e^{\lambda(s-t)} c_s^\alpha \, ds. \tag{2.39}$$

The objective is to obtain

$$V(w, \xi) \equiv \sup E\Big[\int_0^\infty e^{-\rho t} u(\xi_t)\, dt \,\Big| w_0 = w, \xi_0 = \xi \Big]. \tag{2.40}$$

In some sense, we might ideally like to take $\alpha = 1$; this is a degenerate problem, as we will shortly explain. Note however that because of the concave dependence on c in the definition of ξ, we will prefer to have flows of c that are not too bumpy. Despite the fact that there are now two state variables, there is still a nice scaling behaviour which makes it possible to get a one-variable problem. We notice that for any $a > 0$,

$$V(aw, a^\alpha \xi) = a^{(1-R)\alpha} V(w, \xi), \tag{2.41}$$

from which it follows easily that

$$V(w, \xi) = \xi^{1-R} v(w\xi^{-1/\alpha}) \equiv \xi^{1-R} v(z), \tag{2.42}$$

for $v(x) = V(x, 1)$, writing also $z \equiv w/\xi^{1/\alpha}$. The HJB equations for the problem are

$$\sup_{c,\theta} \left[u(\xi) - \rho V + (rw + \theta(\mu - r) - c) V_w + \tfrac{1}{2}\sigma^2\theta^2 V_{ww} + \lambda(c^\alpha - \xi) V_\xi \right] = 0. \tag{2.43}$$

Utilising the scaling property (2.42), writing $\theta = \pi w$ and $c = qw$, a few calculations reduce (2.43) to

$$\sup_{q,\pi} \left[u(1) - \rho v + (r + \pi(\mu - r) - q) z v' + \tfrac{1}{2}\sigma^2\pi^2 z^2 v'' + \lambda(q^\alpha z^\alpha - 1)\{(1-R)v - zv'/\alpha\} \right] = 0. \tag{2.44}$$

The optimal choices[5] of q and π are easily found:

$$\pi = -\frac{(\mu - r)v'}{z\sigma^2 v''}, \tag{2.45}$$

$$q = z^{-1} \left\{ \frac{v'}{\lambda(\alpha(1 - R)v - zv')} \right\}^{1/(\alpha-1)}. \tag{2.46}$$

Inserting these values into the HJB equation, we obtain

$$u(1) - \rho v + rzv' - \frac{\kappa^2 (v')^2}{2v''} - \lambda A + \frac{1-\alpha}{\alpha} (\lambda\alpha A)^{1/(1-\alpha)} (v')^{-\alpha/(1-\alpha)} = 0, \tag{2.47}$$

where

$$A = (1 - R)v - zv'/\alpha.$$

[5] The reason we do not allow $\alpha = 1$ is that the dependence on q is linear, and the problem degenerates; in effect, in this situation it is always possible to transfer an amount of wealth directly into ξ by a delta-function transfer, so the problem is degenerate.

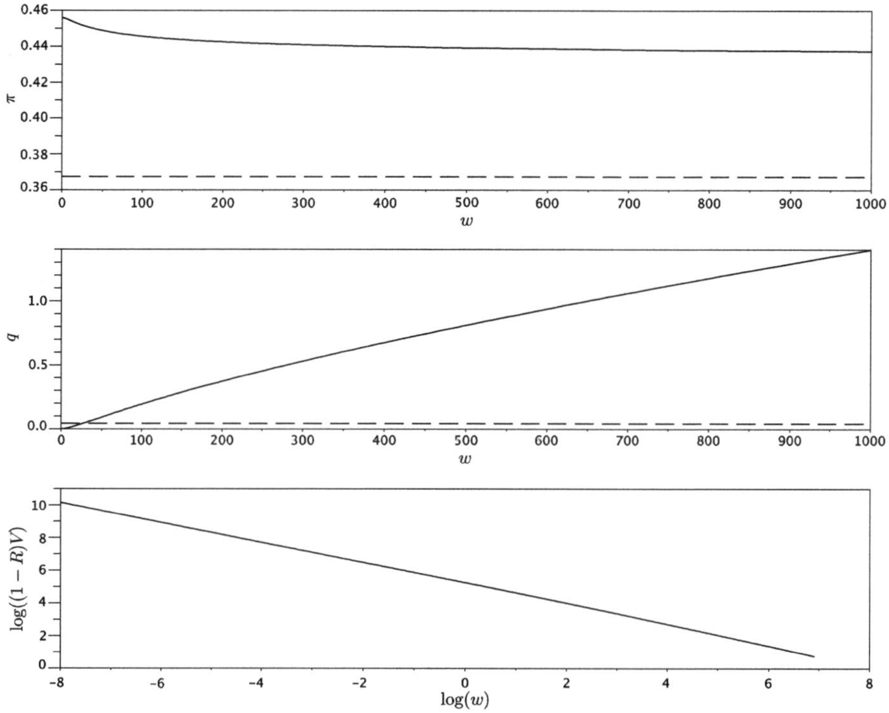

Fig. 2.7 Solving the history-dependent preferences problem of Section 2.7: plots of $\pi = \theta/w$ and $q = c/w$ against w, and of $\log((1 - R)v)$ against $\log w$. These plots use $\alpha = 0.7$, $\lambda = 0.5$, and assume that $\xi = 1$

Numerics. We show in Fig. 2.7 plots of $\pi = \theta/w$, $q = c/w$ and the log of the value for the situation where $\alpha = 0.7$, $\lambda = 0.5$, with ξ held at 1. The plots were calculated using the policy improvement algorithm (see Section 3.6.1). The dashed lines in the top two plots are the solutions to the standard Merton problem for the same parameter values. The proportion of wealth in the risky asset quickly settles down to a value which is a lot higher than for the standard Merton problem. This is not surprising; the investor's preferences in this example are much less fearful of periods when wealth and consumption are low, so we would expect that he will be more risk seeking. The consumption rate gradually rises with increasing wealth, in contrast to the Merton solution. Notice however that the growth is really quite slow. The final plot shows how the value changes with wealth; the log-log plot is quite close to a straight line.

2.8 Non-CRRA Utilities

The use of a CRRA utility is convenient because it allows us to exploit scaling to simplify problems, as we have seen. If we try to solve the Merton problem for u which are not of the usual CRRA form, there are various ways we can proceed. We can use the dual value function approach, from Section 1.3, in particular, we can use the representation (1.53) of the dual value function. It turns out to be computationally and conceptually simpler to work not with the log-Brownian motion Y of (1.52) but with $X_t \equiv \log(Y_t)$, a Brownian motion with constant volatility κ and with drift $m = (\rho - r - \frac{1}{2}\kappa^2)$. In terms of that we may write the dual value function J as (writing $x \equiv \log(y)$)

$$\begin{aligned} J(y) &= E\left[\int_0^\infty e^{-\rho t}\,\tilde{u}(Y_t)\,dt \,\middle|\, Y_0 = y\right] \\ &= E\left[\int_0^\infty e^{-\rho t}\,\tilde{u}(e^{X_t})\,dt \,\middle|\, X_0 = x\right] \\ &= \int r_\rho(x,v)\,\tilde{u}(e^v)\,dv, \end{aligned} \tag{2.48}$$

where $r_\rho(x,v)$ is the resolvent density for X. This needs to be made more precise, and may be expressed in terms of the two roots $\alpha_- < 0 < \alpha_+$ of the quadratic

$$t \mapsto \tfrac{1}{2}\kappa^2 t^2 + mt - \rho$$

as[6]

$$r_\rho(x,v) = r_\rho(v-x) \equiv (m^2+2\rho\kappa^2)^{-1/2}\,\exp\big(\alpha_+ \min\{0, x-v\} + \alpha_- \max\{0, x-v\}\big). \tag{2.49}$$

This allows us to write $J(y)$ from (2.48) as the convolution integral

$$J(y) = \int r_\rho(v-x)\,\tilde{u}(e^v)\,dv. \tag{2.50}$$

If we write

$$\bar{r}_\rho(x) \equiv \int_{-\infty}^{x} r_\rho(v)\,dv,$$

we can perform an integration by parts in (2.50) to express the dual value as

$$J(y) = \big[\bar{r}_\rho(v-x)\tilde{u}(e^v)\big]_{-\infty}^{\infty} + \int \bar{r}_\rho(v-x)\,e^v I(e^v)\,dv, \tag{2.51}$$

exploiting the fact that $\tilde{u}' = -I$. If we have enough control on the behaviour of $\tilde{u}$ at infinity, the evaluation between the limits will vanish, and we are left with a convolution integral solely in terms of the inverse marginal utility I. This may be helpful numerically, because we may be able to specify the utility more easily in terms of I than in terms of $\tilde{u}$.

---

6 We slightly abuse notation here; $r_\rho(x,v)$ is a function of the difference $v-x$ only, so we write $r_\rho(z)$ for $r_\rho(0,z)$.

The representation in (1.53) of the dual value function is not the only way we could approach this problem; we could for example attempt to solve the non-linear ODE (1.48) directly, or we could solve the HJB equation itself by policy improvement. These are entirely workable routes, but they require consideration of appropriate boundary conditions, which may be hard to understand in the case of fairly general choices of the utility. The convolution integral approach we have presented here avoids consideration of the boundary conditions (indeed, they were dealt with on the way to the representation (1.53)), and allows us to reduce the numerics to an application of the Fast Fourier Transform, which is spectacularly efficient.

Numerics. We illustrate the preceding with the example where the inverse marginal utility is

$$I(x) = (x^{1/R_1} + x^{1/R_2})^{-1}, \tag{2.52}$$

where $R_1 = 3$ and $R_2 = 0.8$. For large x, this looks like x^{-1/R_2} and for small x it looks like x^{-1/R_1}, so we will expect that for large wealth the behaviour should be like that of an agent with coefficient of relative risk aversion R_1, and for small wealth the behaviour should be like an agent with coefficient of relative risk aversion R_2. This gives us a utility which tends to zero at zero, and is bounded above. For large wealth, this agent is more risk averse, so we would expect the proportion of wealth he invests in the risky asset to fall with wealth to the Merton proportion for $R = R_1$. His consumption rate should tend to γ_M calculated with $R = R_1$ for large wealth. What do we actually find? The results are plotted in Fig. 2.8. For large wealth, the value function climbs to its asymptotic maximal value. The proportion of wealth invested falls from π_M calculated with $R = R_2$ to the value calculated with $R = R_1$, as expected; the two values of π_M are plotted as dashed lines. We see a similar picture for the consumption rates, but interestingly this rises very slightly above the γ_M value for $R = R_1$ before falling back. This is a genuine feature, not a numerical imprecision; it appears in all the different ways of calculating the solution.

2.9 An Insurance Example with Choice of Premium Level

Here we consider the problem of an insurance company, which is able to invest in a riskless bank account, and a single risky asset, but is also conducting an insurance business, where the volume of business underwritten is determined by the premium charged—the higher the premium, the less business the firm does. For various reasons, it is preferable to treat the volume of business q as the choice variable, and to view the premium rate p as a function of q, that is, $p = p(q)$. The wealth dynamics are taken to be

$$dw_t = rw_t dt + \theta_t(\sigma dW_t + (\mu - r)dt) - c_t dt + q_t p_t dt - dC_t, \tag{2.53}$$

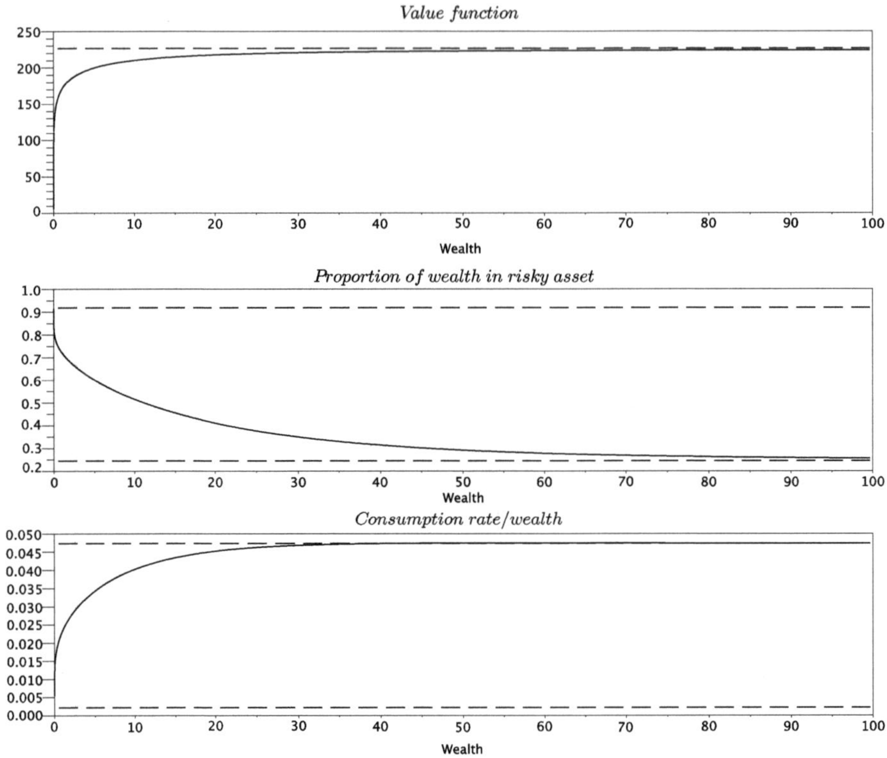

Fig. 2.8 Plots of the value, portfolio and consumption rates for the two-R example of (2.52). For low wealth, behaviour is like an agent with coefficient of relative risk aversion equal to 0.8, and for high wealth the behaviour is like an agent with coefficient of relative risk aversion equal to 3

where C is the total claims process, an increasing compound Poisson process with variable rate q_t. What this means is that $C_t = Y\big(\int_0^t q_s\, ds\big)$, where Y is a compound Poisson process with jumps distributed as F, independent of W:

$$E\exp(-\lambda Y_t) = \exp\left\{-t \int_0^\infty (1-e^{-\lambda x})\; F(dx)\right\}.$$

The consumption rate process c could here be interpreted as a rate of payment of dividends to the shareholders. However we want to understand it, we will propose the objective

$$V(w) \equiv \sup\, E\left[\int_0^\tau e^{-\rho t} u(c_t) dt - K e^{-\rho\tau} \mid w_0 = w\right], \tag{2.54}$$

where K is a penalty for the firm going bankrupt, and τ is the time of bankruptcy, $\tau \equiv \inf\{t : w_t \le 0\}$. The presence of the jumps in C means that there is always a risk that the firm could go broke. Taking the jumps into account, the HJB equation for this problem becomes

$$0 = \sup_{\theta,q,c} \Big[u(c) - \rho V(w) + \{rw + \theta(\mu - r) - c + qp(q)\}V'(w) + \tfrac{1}{2}\sigma^2\theta^2 V''(w)$$
$$+ q\int_0^\infty \{V(w-x) - V(w)\}\, F(dx)\Big]. \tag{2.55}$$

The optimization over q is a novel feature, the optimization over θ and c being as so often before. The other novelty is the integral term arising from the jumps.[7] We will suppose that the utility u is bounded below on $(0, \infty)$, otherwise there may come a time when the bankruptcy penalty may be more desirable than continuing to consume. For an interesting question, then, we shall suppose that $u(0) = 0$, and in the examples studied numerically, we shall have u bounded above as well.

The natural first choice for solving (2.55) is some form of policy improvement, and this can indeed be carried out, with some suitable modifications. We need to modify the method because for a given choice of policy (θ, c, q), the linear system to be solved *will not be sparse*, due to the presence of the integral term in (2.55). If we want to have more than a few hundred grid points, solving a non-sparse system will in general collapse under accumulated errors. So what we do is to generate a sequence V_n of approximations to the value, starting from V_0 being the value we would get if there was no insurance business: $q \equiv 0$. This problem we showed how to solve in Section 2.8. Having found approximation V_n, we generate the next choice of controls by the obvious recipe

$$c_n = I(V_n'), \tag{2.56}$$

$$\theta_n = -(\mu - r)V_n'/\sigma^2 V_n'', \tag{2.57}$$

$$\{q_n p'(q_n) + p(q_n)\}V_n' = \int \{V_n(\cdot) - V_n(\cdot - x)\}\, F(dx). \tag{2.58}$$

Then we find the next approximation to the value function by solving

$$0 = u(c_n) - \rho V(w) + \{rw + \theta_n(\mu - r) - c + q_n p(q_n)\}V'(w) + \tfrac{1}{2}\sigma^2\theta_n^2 V''(w)$$
$$+ q_n \int_0^\infty V_n(w-x)\, F(dx) - q_n V(w). \tag{2.59}$$

Notice particularly that inside the integral there appears the already-known function V_n, so the linear system to be solved for V is sparse. The interpretation of this is that once the first jump occurs, carrying wealth level from w to $w - x$, the remaining reward received is $V_n(w-x)$, which is the best you could have got at the previous level of the recursive solution. It is clear that $V_1 \geq V_0$, because V_0 is the best value which could be achieved if you were not allowed to participate in the insurance market,

[7] Of course, we have $V(w) = -K$ for all $w < 0$.

and V_1 is an improvement, because you are allowed to participate in the insurance market until the first claim. Strictly speaking, we should now hold V_0 fixed inside the integral, and carry out the policy improvement iteration until we have the value for the problem where we are allowed to participate in the insurance market up to the first claim; but it seems in practice to be unnecessary to do this. This illustrates the point that the policy improvement algorithm can work so long as the improvements at each step are indeed improvements; they do not have to be optimal choices.

Numerics. In the numerical example, we used the form $p(q) = q^{-\beta}$ with $\beta = 0.8$, the penalty for bankruptcy was $K = 100$, the claim distribution F was exponential with mean 1, and the utility was the same as used in the example in Section 2.8, with inverse marginal utility (2.52). For boundary conditions, we suppose that for very large wealth the value will be close to the maximal value $V_{\max} = \sup_x u(x)/\rho$, and will be assumed to have the form

$$V(w) = V_{\max} + A(u(w) - u(\infty)) \tag{2.60}$$

for some $A > 0$. For zero wealth, there will be no investment in the risky asset, as this would immediately bankrupt the firm. Instead, we find ourselves looking at the condition

$$0 = \sup_{q,c} \left[u(c) - \rho V(0) + \{qp(q) - c\}V'(0) - q(K + V(0)) \right] \tag{2.61}$$

which gives the boundary condition at 0.

The plots in Fig. 2.9 show what happened. The value is visibly higher than the value for the problem with no access to the insurance market, shown as a dashed line. Both lines asymptote to the maximal value $V_{\max}$, but a separate plot (Fig. 2.10) of the relative value

$$\frac{V_{\max} - V(\cdot)}{V_{\max} - V_0(\cdot)}$$

is steadily increasing, showing that the advantage of having access to the insurance market continues to grow as the firm gets more wealthy, as would be expected, since the risk of default recedes. The optimal value of business falls with wealth, but not to zero; when wealth is zero, it will be optimal to invest nothing in the risky asset, but there is an incentive nevertheless to invest in the insurance business, since there will be premium income before the first claim, and this boosts the growth of wealth and consumption. The level of business starts slowly, but levels off to an asymptotic value. Looking at the wealth in the risky asset, and the consumption rate, we see that these are always both greater than the corresponding values for the problem with no insurance business, and that the surge in the increase happens for the same wealth values where the volume of business surges, all growing quite rapidly between the values $w = 6$ and $w = 15$. The price charged, shown in Fig. 2.10, falls rapidly as the size of the firm grows.

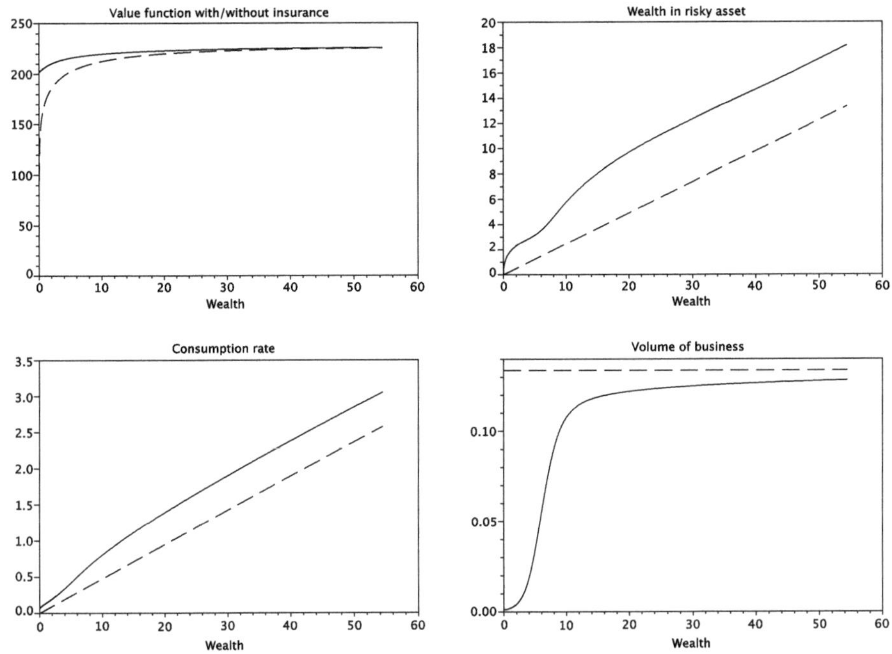

Fig. 2.9 Plots of the value, portfolio, consumption rate and level of business for the insurance problem

2.10 Markov-Modulated Asset Dynamics

Here we suppose that there is some Markov chain ξ taking values in the finite set $I = \{1, 2, \ldots, N\}$, which is independent of the driving Brownian motion W. We let Q denote the $N \times N$ matrix of jump intensities. The volatility and the growth rate of the stock depend on the value of ξ, so that the dynamics of the single risky asset become

$$dS_t/S_t = \sigma(\xi_t)dW_t + \mu(\xi_t)dt \tag{2.62}$$

for some functions σ, μ of the chain, and the wealth dynamics become

$$dw_t = rw_t dt + \theta_t \sigma(\xi_t)(dW_t + \kappa(\xi_t)dt) - c_t dt, \tag{2.63}$$

where $\kappa(\xi) = \sigma(\xi)^{-1}(\mu(\xi) - r)$ is the market price of risk. There are two[8] radically different situations to be dealt with:

[8] Combinations of the two cases could be considered, where the function σ takes more than 1 value, but fewer than N. This could be handled by similar techniques, but we omit discussion as it is not particularly relevant.

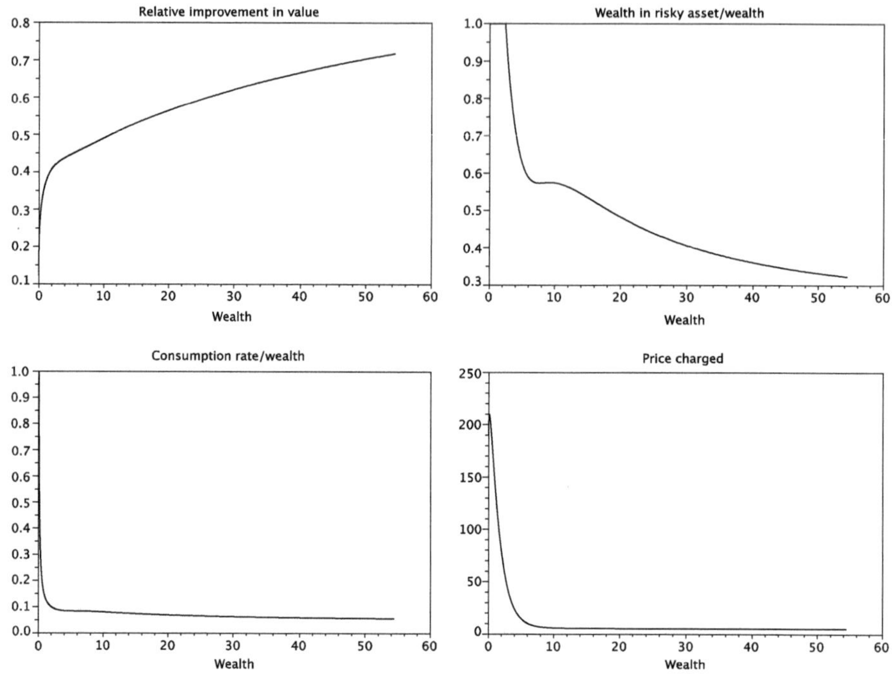

Fig. 2.10 Plots of the relative value, portfolio proportion, consumption rate divided by wealth, and price charged for the insurance problem of Section 2.9

1. The function σ is one-to-one;
2. The function σ is constant.

In the first situation, by observing the quadratic variation of the stock, we can deduce the value of ξ; in the second, the value of ξ has to be filtered from the observations. The treatment of the second situation is more complicated, but we can deal with both.

Case 1: ξ is observed. The value of the problem depends on ξ as well as on w, so the value function

$$V(w,\xi) \equiv \sup E\Big[\int_0^\infty e^{-\rho s} u(c_s)\, ds \,\Big|\, w_0 = w, \xi_0 = \xi\Big]$$

will satisfy the HJB equations[9]

$$0 = \sup_{\theta,c} \big[u(c) - \rho V + \tfrac{1}{2}\theta^2\sigma^2 V_{ww} + (rw - c + \theta(\mu - r))V_w + QV \big].$$

[9] We use the notation QV as a shorthand for the function $(QV)(w,\xi)$ defined to be $(QV)(x,\xi) \equiv \sum_{j\in I} q_{\xi j} V(x,j)$.

From scaling it is clear that $V(w,\xi) = u(w)f(\xi)$, so after substituting this form of V and simplifying we learn that

$$0 = Rf^{1-1/R} - \{\rho + (R-1)(r + \tfrac{1}{2}\kappa^2/R)\}f + Qf. \tag{2.64}$$

Numerical solution of (2.64) is relatively simple; we just recursively solve the linear equations

$$\{\rho + (R-1)(r + \tfrac{1}{2}\kappa^2/R)\}f_n - Qf_n = Rf_{n-1}^{1-1/R}$$

from some suitable positive starting point f_0, and this is very quick.

Case 2: ξ has to be filtered. This is the situation where the volatility σ is constant, so that the volatility of the stock price does not reveal the underlying Markovian state. Let us write

$$dY_t \equiv dW_t + \kappa(\xi_t)dt, \tag{2.65}$$

which is observable.[10] Let $(\mathscr{Y}_t)_{t\geq 0}$ be the (usual augmentation[11] of) the filtration generated by the process Y. The wealth dynamics can now be expressed in the form

$$\begin{aligned} dw_t &= rw_t dt + \theta_t \sigma dY_t - c_t dt & (2.66)\\ &= rw_t dt + \theta_t \sigma (dN_t + \hat{\kappa}_t) - c_t dt, & (2.67) \end{aligned}$$

where $\hat{\kappa}$ is the $\mathscr{Y}$-optional projection of the process $\kappa_t \equiv \kappa(\xi_t)$, and N is the *innovations process*, a $\mathscr{Y}$-Brownian motion. This is a familiar story from filtering theory; see, for example, [34], VI.8 for more background.

If we write $\pi_t(x) = P(\xi_t = x|\mathscr{Y}_t)$, $x \in I$, for the posterior of ξ given the observations, then the evolution of π is given[12] by the system of equations:

$$d\pi_t(x) = \pi_t(x)(\kappa(x) - \hat{\kappa}_t)dN_t + (Q^T\pi_t)(x)dt, \qquad (x \in I). \tag{2.68}$$

Now (2.66) and (2.68) together form an $(N+1)$-dimensional SDE driven by Y, or equivalently, N, and this can in principle be solved.[13]

Let us now specialize to the case of $N = 2$, so that $I = \{1, 2\}$ and

$$Q = \begin{pmatrix} -\alpha & \alpha \\ \beta & -\beta \end{pmatrix}.$$

We write $p_t \equiv \pi_t(1) = 1 - \pi_t(2)$. In terms of this, we have

$$\hat{\kappa}_t = p_t\kappa_1 + (1-p_t)\kappa_2,$$

---

10 The process Y is the log of the discounted asset, divided by σ.

11 See [33], II.67.

12 See [34], VI.11.

13 Notice that $\hat{\kappa}_t = \langle \pi_t, \kappa \rangle$, so that the drift in dY is expressed in terms of π_t.

and we have the coupled equations

$$dw_t = rw_t\,dt + \theta\sigma(dN_t + \hat{\kappa}_t) - c_t\,dt, \tag{2.69}$$
$$dp_t = p_t(\kappa_1 - \hat{\kappa}_t)dN_t + \{\beta(1-p_t) - \alpha\}dt. \tag{2.70}$$

Now the value function for this problem is a function of both w and p

$$V(w,p) \equiv \sup_{(n,c)\in\mathscr{A}(w)} E\left[\int_0^\infty e^{-\rho t}u(c_t)\,dt\,\middle|\,w_0 = w,\,p_0 = p\right],$$

satisfying the HJB equations

$$\begin{aligned}0 = \sup_{c,\theta}\big[&u(c) - \rho V(w,p) + \{rw + \theta\sigma(p\kappa_1 + (1-p)\kappa_2) - c\}V_w(w,p)\\ &+ \tfrac{1}{2}\theta^2\sigma^2 V_{ww}(w,p) + \sigma\theta p(1-p)(\kappa_1-\kappa_2)V_{wp}(w,p)\\ &+ \tfrac{1}{2}p^2(1-p)^2(\kappa_1-\kappa_2)^2 V_{pp}(w,p)\big].\end{aligned}$$

Optimizing over c and θ gives

$$\begin{aligned}c &= V_w^{-1/R},\\ \sigma\theta V_{ww} &= -p(1-p)(\kappa_1-\kappa_2) - (p\kappa_1 + (1-p)\kappa_2).\end{aligned}$$

As usual, for CRRA u we deduce the scaling relation $V(w,p) = u(w)f(p)$; substituting this back into the HJB equations yields

$$\begin{aligned}0 = &Rf^{1-1/R} - \rho f + r(1-R)f + (\beta - (\alpha+\beta)p)f' + \tfrac{1}{2}p^2(1-p)^2(\kappa_1-\kappa_2)^2 f''\\ &+ (1-R)\left\{p(1-p)(\kappa_1-\kappa_2)f' + (p\kappa_1 + (1-p)\kappa_2)f\right\}^2/2Rf \end{aligned}\tag{2.71}$$

after some simplifications. This is easily solved by policy improvement, or more simply by iterative solution, as explained in Section 3.6.2.

Numerics. A numerical example has been calculated using $\alpha = 0.15$, $\beta = 0.20$, $\mu(1) = 0.07$ and $\mu(2) = 0.17$, and the results are shown in Fig. 2.11. The horizontal axis in each plot is the posterior probability of being in state 1, the low-growth state. As this rises, we see that the efficiency, consumption rate and proportion of wealth invested in the risky asset all decrease, as would be expected. The efficiency drops from 1.03 to 0.97, which is relatively insubstantial, as is the fall in the consumption rate. However, the proportion of wealth invested in the risky asset falls from 49 % to 7 %, a very substantial reduction. The relative insensitivity of the efficiency and consumption rate to the posterior probability of being in the low-growth state is to some extent to be explained by the fact that with $\rho = 0.02$ the agent has a very long horizon, of mean 50 years, whereas the state of the chain is switching every 5–6 years on average. Thus the effect which this patient agent sees will be quite like the average value; he will not reduce or expand consumption much from the Merton

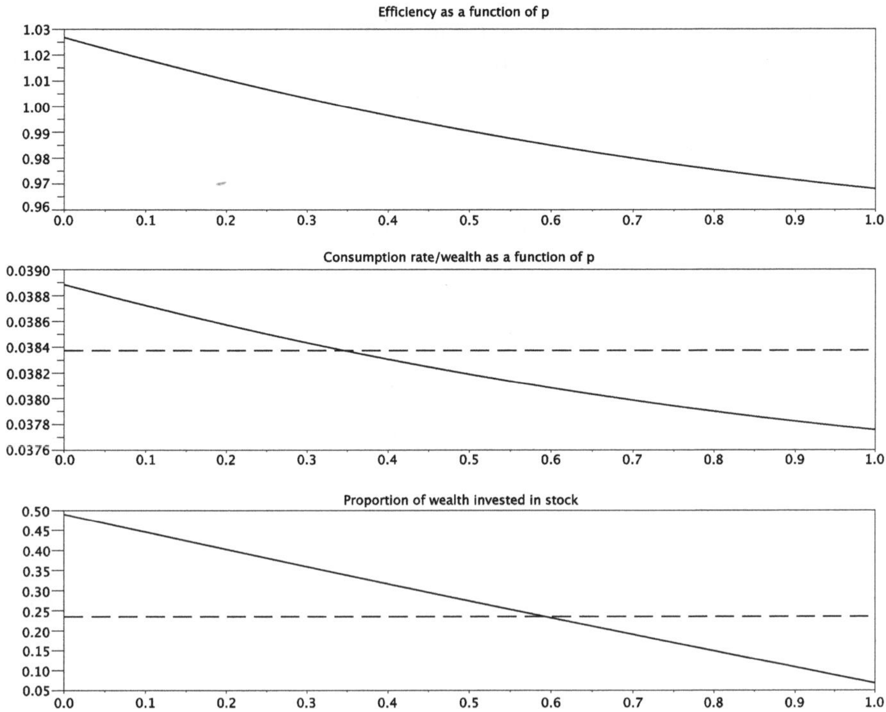

Fig. 2.11 Plots of efficiency, consumption rate and proportion of wealth in risky asset for the model of Section 2.10, compared to the values for the standard Merton problem where the growth rate μ is constant and equal to the mean of the growth rate of the hidden Markov chain

values for the average growth rate, because over the timescales he cares about bad times and good times will even out. Nevertheless, he varies his investment mix quite substantially as the posterior probability moves, to take advantage of whichever asset, the stock or the bank account, is better for him at any given time.

2.11 Random Lifetime

Suppose that an agent lives for a random time τ which is independent of the evolution of the assets, and has a distribution specified in terms of its (deterministic) hazard rate $h : \mathbb{R}^+ \to \mathbb{R}^+$ by

$$P[\tau > t] = \exp\Big(-\int_0^t h(s)\,ds\Big) \qquad (t > 0). \tag{2.72}$$

The agent's objective is to maximize

$$E\left[\int_0^\tau \varphi(s)u(c_s)\,ds\right] \tag{2.73}$$

where φ is some deterministic function which reflects the agent's preferences over the different times of consumption; for example, it may be that the agent cares more about consumption in his old age. What is the agent's optimal behaviour?

Assuming as we often do that $u'(x) = x^{-R}$ for some positive R different from 1, we have that the value function

$$V(t, w) \equiv \sup E\left[\int_t^\tau \varphi(s) u(c_s)\, ds \,\middle|\, w_t = w, \tau > t \right] \tag{2.74}$$

will have the familiar scaling form

$$V(t, w) = f(t) u(w) \tag{2.75}$$

for some function f which is to be found. For this problem, the HJB equation is

$$\begin{aligned}
0 &= \sup_{c,\theta} \left[\varphi\, u(c) + V_t + (rw + \theta(\mu - r) - c) V_w + \tfrac{1}{2}\sigma^2\theta^2 V_{ww} - hV \right] \\
&= \sup_{x,z} \left[\varphi\, u(wx) + u(w) f' + w(r + z(\mu - r) - x) V_w + \tfrac{1}{2}\sigma^2 z^2 w^2 V_{ww} - hV \right] \\
&= \sup_{x,z} \left[\varphi\, u(wx) + u(w) f' + (1 - R) f (r + z(\mu - r) - x) u(w) - \tfrac{1}{2}\sigma^2 z^2 R(1 - R) f u(w) - hf u(w) \right] \\
&= \sup_{x,z} u(w) \left[\varphi\, x^{1-R} + f' + (1 - R) f (r + z(\mu - r) - x) - \tfrac{1}{2}\sigma^2 z^2 R(1 - R) f - hf \right]
\end{aligned} \tag{2.76}$$

where we have written $\theta = wz$, $c = wx$ in the development. Now the optimization over x and z is easy to do, and we find optimal values

$$z^* = \frac{\mu - r}{\sigma^2 R} \equiv \pi_M, \qquad x^* = \left(\frac{\varphi(t)}{f(t)} \right)^{1/R}. \tag{2.77}$$

The message therefore is that we invest according to the Merton proportion, but the consumption rate is *not* a constant times the wealth, but depends on time in a deterministic way. The form of the optimal solution is hardly surprising, but we can offer more than just some verbal description of the form of the solution; we can in fact find the optimal solution, by solving the HJB equation for f, which here is a non-linear first-order ODE:

$$0 = f' - (h + (R - 1)(r + \kappa^2/2R)) f + R\varphi^{1/R} f^{1-1/R}, \tag{2.78}$$

as we find by substituting back the values (2.77) into (2.76).

Remarkably, some well-chosen substitutions reduce the ODE (2.78) to a much simpler ODE which we can solve. If we set $b \equiv (R - 1)(r + \kappa^2/2R)$, and

$$\psi(t) = \exp\left(-bt - \int_0^t h(s)\, ds \right)$$

then $g(t) \equiv f(t)\psi(t)$ is easily seen to solve

$$g'(t) + \tilde{\varphi}(t)g(t)^{1-1/R} = 0, \tag{2.79}$$

where

$$\tilde{\varphi}(t) \equiv R(\varphi(t)\psi(t))^{1/R}$$

is a known function. Thus

$$\frac{d}{dt}\left[g(t)^{1/R}\right] = -\frac{\tilde{\varphi}(t)}{R}. \tag{2.80}$$

All we need to solve this is some boundary condition; probably the simplest thing to do is to assume that $\varphi(t) = 0$ for all $t \geq T_0$ for some fixed $T_0 > 0$, which then fixes $f(T_0) = 0$, and so

$$g(t)^{1/R} = \int_t^{T_0} \frac{\tilde{\varphi}(s)}{R}\, ds. \tag{2.81}$$

2.12 Random Growth Rate

This example is quite similar to the example in Section 2.2 where the interest rate is not assumed to be constant, but evolves as an OU model. Here we take the wealth dynamics to be

$$\begin{aligned} dw_t &= rw_t dt + \theta(\sigma dW_t + (\mu_t - r)dt) - c_t dt \\ d\mu_t &= \sigma_\mu dB_t + \beta(\bar{\mu} - \mu_t)\, dt, \end{aligned}$$

where now the growth rate is no longer supposed constant, but follows an OU process. The parameters σ_μ and $\bar{\mu}$ are constants; let us suppose that the two Brownian motions W and B are correlated, $dBdW = \eta dt$. The objective of the agent is to obtain

$$V(w, \mu) = \sup E\left[\int_0^\infty e^{-\rho t} u(c_t)\, dt\, \middle|\, w_0 = 0, \mu_0 = \mu\right] \tag{2.82}$$

where as usual $u(w) = w^{1-R}/(1-R)$. A moment's reflection shows that the solution of the Merton problem will still scale, with the value function taking the form

$$V(w, r) = u(w)g(\mu).$$

Seeking the HJB equation for this problem, we find (substituting $q = c/w, s = \theta/w$)

$$\begin{aligned}0 &= \sup\big[u(c) - \rho V + \tfrac{1}{2}\sigma^2\theta^2 V_{ww} + \eta\sigma\sigma_\mu\theta V_{w\mu} \\ &\quad + \tfrac{1}{2}\sigma_\mu^2 V_{\mu\mu} + (rw + \theta(\mu - r) - c)V_w + \beta(\bar{r} - r)V_r\big] \\ &= \sup u(w)\big[q^{1-R} - q(1-R)g - \rho g - \tfrac{1}{2}R(1-R)\sigma^2 s^2 g + (1-R)\eta\sigma\sigma_\mu s g' + \tfrac{1}{2}\sigma_\mu^2 g'' \\ &\quad + (r + s(\mu - r))(1-R)g + \beta(\bar{\mu} - \mu)g'\big].\end{aligned}$$

Now optimising this over q and s gives us

$$\begin{aligned}q &= g^{-1/R}, \\ s &= \frac{\eta\sigma\sigma_\mu g' + (\mu - r)g}{\sigma^2 R g},\end{aligned}$$

and when substituted back in gives the following second-order ODE for the HJB equations:

$$0 = Rg^{1-1/R} - \rho g + r(1-R)g + (1-R)\,\frac{(\eta\sigma\sigma_\mu g' + (\mu - r)g)^2}{2R\sigma^2 g} + \tfrac{1}{2}\sigma_\mu^2 g'' + \beta(\bar{\mu} - \mu)g'. \tag{2.83}$$

Here, $\kappa = (\mu - r)/\sigma^2 R$. As before, Eq. (2.83) cannot be solved in closed form, but the numerical solution is not particularly difficult.

It is worth comparing the HJB equation (2.83) obtained here with the HJB equation (2.13) obtained in Section 2.2 for the example with stochastic interest rate. At first glance, apart from trivial notational switches, they appear to be identical. But they are not; in (2.83) μ is a *variable* and r is a *constant*, and in (2.13) it is the other way round.

Notice that the problem considered here is rather unrealistic; we would not in practice know what the value of μ is, so the solution is academic. A more interesting version of this problem is treated in Section 2.27, where we have to estimate μ from the observed prices.

Numerics. We show in Fig. 2.12 plots of efficiency, consumption rate $q = c/w$ and portfolio proportion $s = \theta/w$ for an example where we took $\sigma_\mu = 0.05$, $\bar{\mu} = 0.14$, $\eta = 0.6$, and $\beta = 0.5$. In the plot of efficiency, the level 1 is marked with a dashed line; in the plot of consumption rate, the dashed line shows the consumption rate that would hold in the Merton problem where μ was constant and equal to μ_0; and in the plot of portfolio proportion, the dashed line shows the fraction of wealth to be invested in the risky asset if μ were constant and equal to μ_0. The problem was solved numerically using policy improvement with reflecting boundary conditions at the end of a wide interval containing the region plotted. Unsurprisingly, efficiency rises as μ_0 rises, once μ_0 gets far enough away from 0. The plot of the proportions invested in the risky asset shows little difference in what is optimal and what would be optimal for the Merton problem with constant μ. However, the plot of the consumption rate shows substantial differences; if the growth rate is high and is constant, then we will want to consume rapidly, because we expect to get good returns for ever, but if the

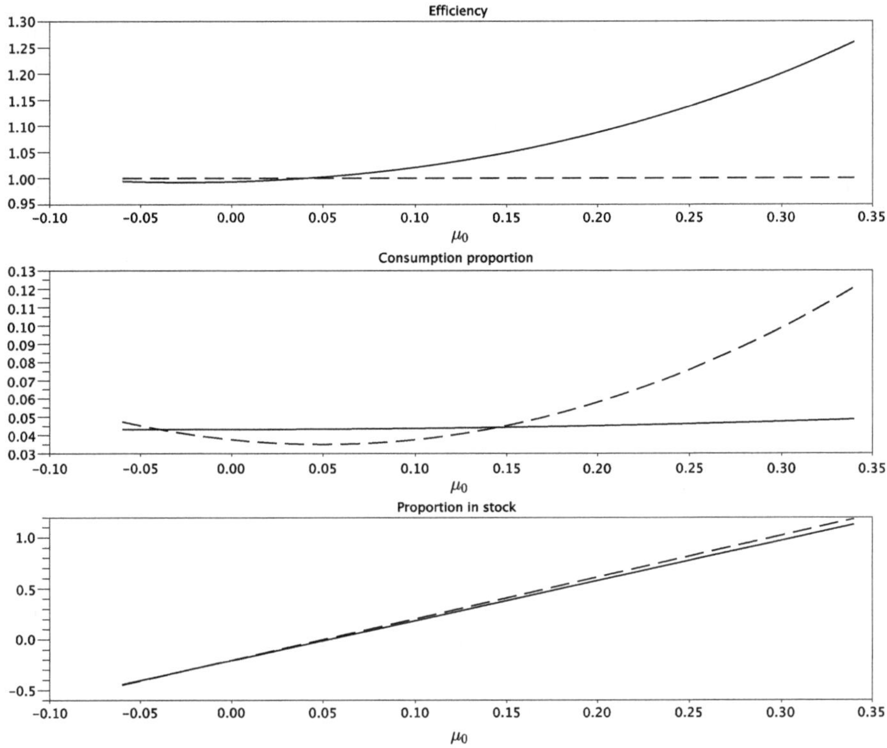

Fig. 2.12 Plots of efficiency, consumption rate, and proportion of wealth in the risky asset for the problem of Section 2.12 with randomly-varying growth rate. Parameter values were $\sigma_\mu = 0.05$, $\bar{\mu} = 0.14$, $\eta = 0.6$, and $\beta = 0.5$

growth rate is high and random, we are more cautious, since the growth rate will soon fall back to more normal levels.

Different choices of the parameters σ_μ, $\bar{\mu}$, η and β can produce quite different plots. For example, changing η to -0.1 gives efficiencies in excess of 1 everywhere. Making $\beta = 0.15$ again leads to efficiencies in excess of 1 everywhere, but not by so much. Changing σ_μ to 0.15 leads to efficiencies in excess of 1.65 everywhere, a striking difference!

2.13 Utility from Wealth and Consumption

Here we shall once again assume standard wealth dynamics (2.1) but that the objective of the agent is

$$V(w) \equiv \sup E\left[\int_0^\infty e^{-\rho t} u(c_t, w_t)\, dt \,\middle|\, w_0 = w\right]. \tag{2.84}$$

We could arrive at such an objective if we wanted to model the phenomenon that consuming more makes an agent happier, but if his rate of consumption is too large a fraction of his current wealth, then the happiness is diminished. The HJB equation for this problem is by now easy to write down:

$$0 = \sup_{c \geq 0, \theta} \left[-\rho V + u(c, w) + \{rw + \theta(\mu - r) - c\} V_w + \tfrac{1}{2}\sigma^2\theta^2 V_{ww} \right]. \tag{2.85}$$

With the notation $\tilde{u}(y, w) = \sup_c \{u(c, w) - yc\}$ we can perform the optimizations over c and θ to obtain

$$0 = -\rho V + \tilde{u}(V_w, w) + rwV_w - \tfrac{1}{2}\kappa^2 \frac{V_w^2}{V_{ww}}. \tag{2.86}$$

Again, without scaling properties it is hard to advance further. But if we assume that

$$u(c, w) = \frac{w^\alpha c^\beta}{1 - R} \tag{2.87}$$

for some α, β of the same sign as $1 - R$, $\alpha + \beta = 1 - R$, then scaling gives us that $V(\lambda w) = \lambda^{1-R} V(w)$ for all $\lambda > 0$, and hence

$$V(w) = Au(w)$$

for some positive A, where $u(w) = w^{1-R}/(1 - R)$. Substituting this form into (2.85) we find that

$$0 = -AR\gamma_M + \left(\frac{(1 - R)A}{\beta} \right)^{\beta/(\beta-1)} (1 - \beta).$$

Rearranging gives us that

$$A^{1/(\beta-1)} = \left(\frac{\beta}{1 - R} \right)^{-\beta/(\beta - 1)} \frac{R\gamma_M}{1 - \beta}. \tag{2.88}$$

As might have been anticipated, we find that the optimal investment rule is just the Merton rule, and that we consume proportionally to wealth, though the constant of proportionality is in general not γ_M. As a check, we must find that if $\alpha = 0$ we recover the solution to the original Merton problem; indeed, in this case we have $\beta = 1 - R$, and the expression (2.88) tallies with the original Merton solution (1.9).

2.14 Wealth Preservation Constraint

In this version of the Merton problem, the wealth dynamics are the standard ones (2.1), but we shall impose the constraint that the wealth of the agent is *preserved*, in the sense that

$$w_t \geq b\bar{w}_t \equiv b \int_{-\infty}^{t} \lambda e^{\lambda(s-t)} w_s \, ds, \tag{2.89}$$

where $b \in (0, 1)$ is a constant, as is $\lambda > 0$. We make the convention that $w_s = w_0$ for all $s < 0$. This models the notion that we will not want our wealth to fall too much below what it has been in the past, as represented by the exponentially-weighted moving average $\bar{w}_t$. The dynamics of $\bar{w}$ are given by

$$d\bar{w} = \lambda(w - \bar{w})dt. \tag{2.90}$$

The objective of the agent is to obtain

$$V(w, \bar{w}) \equiv \sup E\left[\int_0^\infty e^{-\rho t} u(c_t) \, dt \,\middle|\, w_0 = w, \bar{w}_0 = \bar{w} \right]. \tag{2.91}$$

Using the dynamics (2.1) and (2.90), the HJB equations can be written down:

$$0 = \sup_{c \geq 0, \theta} \left[-\rho V + u(c) + \{rw + \theta(\mu - r) - c\} V_w + \tfrac{1}{2}\sigma^2\theta^2 V_{ww} + \lambda(w - \bar{w}) V_{\bar{w}}. \right], \tag{2.92}$$

As so often, there is little we can do here without some scaling assumptions, so if we assume that u is CRRA, $u'(x) = x^{-R}$ for some $R > 0$ different from 1, then we have the scaling relation

$$V(w, \bar{w}) = \bar{w}^{1-R} v(x) \equiv \bar{w}^{1-R} v(w/\bar{w}). \tag{2.93}$$

Exploiting this form of V in (2.92) leads to the form

$$\begin{aligned} 0 &= \sup_{s \geq 0, q} \left[-\rho v + u(s) + (rx + q(\mu - r) - s)v' + \tfrac{1}{2}\sigma^2 q^2 v'' + \lambda(x - 1)((1 - R)v - xv') \right] \\ &= -\rho v + \tilde{u}(v') + rxv' - \tfrac{1}{2}\kappa^2 \frac{v'^2}{v''} + \lambda(x - 1)((1 - R)v - xv'). \end{aligned} \tag{2.94}$$

Once again, there appears to be no prospect of solving this except numerically.

Numerics. At the lower boundary $x \equiv w/\bar{w} = b$, it has to be that the agent comes out of the risky asset entirely, because the right-hand side of (2.89) is differentiable, whereas the left-hand side will have quadratic variation if there is non-zero holding of the risky asset, and the inequality will be immediately violated. Moreover, we must insist that the consumption rate is not so large that the drift of $w - b\bar{w}$ is negative.

For very large values of x, the dominant effect is that the exponentially-weighted mean $\bar{w}$ is rising very fast, so x is falling very fast. We shall impose the boundary condition that to the right of some suitably large x^* the agent is not allowed to invest in the risky asset. As can be found by varying x^*, this makes almost no difference to the solution even when x^* is not very big.

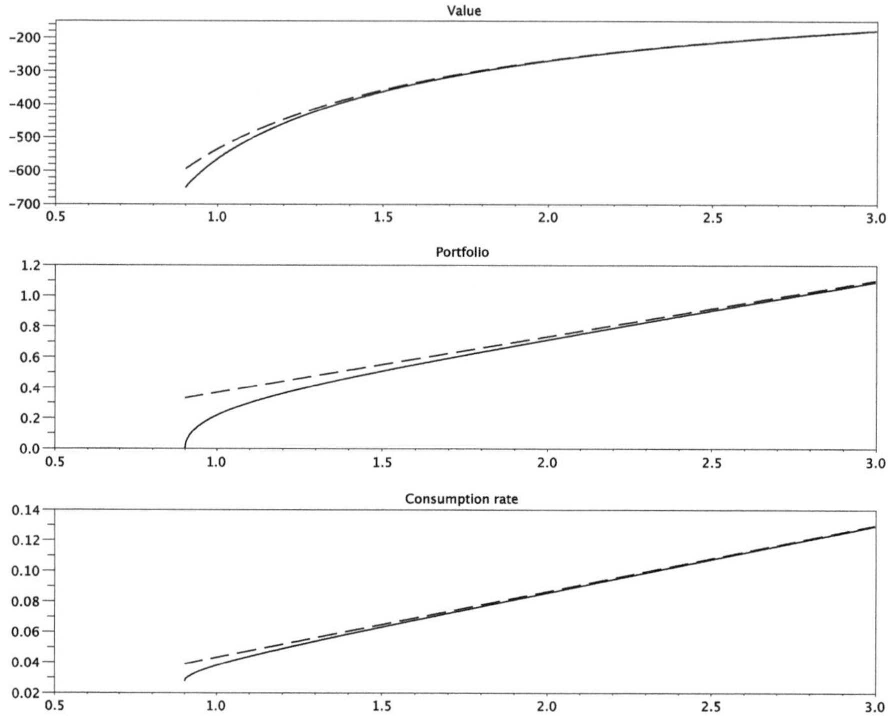

Fig. 2.13 Plots of the value, portfolio and consumption rates for the wealth preservation example of Section 2.14. The Merton solution is shown as *dashed lines*. Values used were $b = 0.9, \lambda = 0.01$

With $b = 0.9$ and $\lambda = 0.01$, and supposing that $w_0 = \bar{w}_0 = 1$, the efficiency is 0.9479. The plots in Fig. 2.13 show the value, portfolio and consumption rates as functions of x. The value lies everywhere below the Merton value, as would be expected, and we see that the effect on consumption is relatively small. The effect on the portfolio is also quite localized; at the critical value b the portfolio of course drops down to zero, but it climbs quite quickly back to the Merton solution as x rises. Overall, then, the effect of this restriction on the agent's behaviour is small, even when the small value of λ means that the lower barrier moves quite slowly. This is probably explained by the fact that the wealth of the Merton investor is growing at rate $(r + \kappa^2(1 + R^{-1})/2 - \rho)/R$ which for the default values is positive. Thus the wealth process is moving away from its historical values generally, so the constraint that w should not fall below $b\bar{w}$ is unlikely to bite often.

2.15 Constraint on Drawdown of Consumption

This is a problem solved by Arun Thillaisundaram [1]. The wealth dynamics are the standard wealth dynamics (2.1), but we now insist that there is limited drawdown of consumption rate:

$$c_t \geq b\bar{c}_t \equiv b \sup_{u \leq t} c_u \tag{2.95}$$

for some constant $b \in [0, 1)$. Otherwise, the agent has the standard objective (2.2), and seeks to obtain the value

$$V(w, \bar{c}) = \sup_{c,\theta} E\left[\int_0^\infty e^{-\rho t} u(c_t)\, dt \,\middle|\, w_0 = w, \bar{c}_0 = \bar{c} \right]. \tag{2.96}$$

Using the Martingale Principle of Optimal Control, we find the HJB equations

$$0 \geq \sup_{c \geq bw, \theta} \left[-\rho V + u(c) + (rw + \theta(\mu - r) - c)V_w + \tfrac{1}{2}\sigma^2\theta^2 V_{ww} \right] \tag{2.97}$$

$$0 \geq V_{\bar{c}}, \tag{2.98}$$

with at least one of the inequalities holding with equality at each x. This implicitly assumes that the maximal consumption rate $\bar{c}$ will only get increased on a set of zero Lebesgue measure, as is typical of a local time. This hypothesis needs to be substantiated by a proper verification argument, but is correct.

Assuming a CRRA felicity $u'(x) = x^{-R}$ allows us to make a scaling and express

$$V(w, \bar{c}) = \bar{c}^{1-R} V(w/\bar{c}, 1) \equiv \bar{c}^{1-R} v(w/\bar{c}) \equiv \bar{c}^{1-R} v(x). \tag{2.99}$$

We expect that if wealth w is large enough relative to $\bar{c}$, then it will make sense to raise $\bar{c}$, but otherwise we do not. So this leads us to suspect that there will be some critical value x^* of $x \equiv w/\bar{c}$ such that when $x > x^*$ we will raise $\bar{c}$ to move x down to x^*. By inspection of the scaling relation (2.99), this tells us that to the right of x^* we must have $v(x) \propto x^{1-R}$, that is, $v(x) = Au(x)$ to the right of x^* for some positive A.

Another feature of the solution is that there is a minimal possible level of wealth consistent with maintaining consumption at the level $b\bar{c}$; indeed, if wealth falls to $b\bar{c}/r$, then we must put all our money into the bank account, and consume the interest, which is paid at rate $b\bar{c}$. If we do that, then the value of the objective will be $u(b)/\rho$. Thus we have determined that

$$v(b/r) = u(b)/\rho. \tag{2.100}$$

Using the scaling relation (2.99) again, the second condition (2.98) is now simply the condition

$$0 \geq (1 - R)v(x) - xv'(x). \tag{2.101}$$

The first condition (2.97) needs a bit more development. Using the scaling relation, we obtain

$$
\begin{aligned}
0 &\geq -\rho v + rxv' - \tfrac{1}{2}\kappa^2 \frac{(v')^2}{v''} + \sup_{b \leq z \leq 1} [u(z) - zv'] \\
&= -\rho v + rxv' - \tfrac{1}{2}\kappa^2 \frac{(v')^2}{v''} + \tilde{u}_b(v'),
\end{aligned} \tag{2.102}
$$

where we define

$$
\begin{aligned}
\tilde{u}_b(z) &\equiv \sup_{b \leq y \leq 1} [u(y) - yz] \\
&= (u(1) - z) I_{\{z < u'(1)\}} + \tilde{u}(z) I_{\{u'(1) \leq z \leq u'(b)\}} + (u(b) - bz) I_{\{u'(b) < z\}}.
\end{aligned} \tag{2.103}
$$

This of course invites us to use the dual variable $z = v'(x)$, with $J(z) = v(x) - xz$, converting the non-linear ODE into the linear dual ODE

$$
0 \geq \tfrac{1}{2}\kappa^2 z^2 J'' + (\rho - r) z J' - \rho J + \tilde{u}_b(z). \tag{2.104}
$$

The condition (2.101) converts to $(1 - 1/R)J - zJ' \geq 0$.

Solving the dual HJB Equation (2.104) with equality gives the solution J as

$$
J(z) = \begin{cases} u(1)/\rho - z/r + A_2 z^{-\alpha} + B_2 z^{\beta} & (z \leq u'(1) = 1) \\ \gamma_M^{-1} \tilde{u}(z) + A_1 z^{-\alpha} + B_1 z^{\beta} & (1 \leq z \leq u'(b)) \\ u(b)/\rho - bz/r + A_0 z^{-\alpha} & (u'(b) \leq z) \end{cases}.
$$

Here we have in each interval found a particular solution, and added a general solution to the homogeneous ODE to get this form. Notice that in $(u'(b), \infty)$ there can be no term of the form z^{β}, because $\beta > 1$, and such a term would either destroy convexity of J, or monotonicity. Moreover, the coefficient A_0 must be non-negative for the solution to be convex.

Once A_0 is determined, we deduce A_1, B_1, A_2, B_2 from the C^1 condition at $u'(b)$ and at $u'(1)$, so the solution J is determined up to the constant A_0. To solve this, what we can do is to work out what the C^1 solution $g(z)$ would be if we took $A_0 = 0$, and then of course $J(z) = g(z) + A_0 z^{-\alpha}$.

There are two further requirements. One is to make A_0 (and hence v) as large as possible; and the other is that at some z we must find that

$$
\left(1 - \frac{1}{R}\right) J(z) = zJ'(z), \tag{2.105}
$$

because at this place we pass from the piecewise-defined solution above to some multiple of $\tilde{u}(z)$, corresponding to the observation that for $x \geq x^*$ the value has the form $v(x) = Au(x)$. Since $J(z) = g(z) + A_0 z^{-\alpha}$, what (2.105) says is

$$
\left(1 - \frac{1}{R}\right)(g(z) + A_0 z^{-\alpha}) = z\{g'(z) - \alpha A_0 z^{-1-\alpha}\}.
$$

Rearranging gives

$$A_0 = \frac{zg'(z) - (1 - 1/R)g(z)}{1 + \alpha - 1/R}. \tag{2.106}$$

Now we just maximize the right-hand side over z to find z^* and A_0, and this gives the entire solution.

Numerics. The only parameter to be specified in addition to the default values (2.3) is the parameter b, taken in this numerical study to be equal to 0.7. The plots in Fig. 2.14 show the value function v, the consumption, and portfolio as a function of the state variable $x = w/\bar{c}$, and as a check the two operators applied to v, the HJB operator (2.102) and the first-order operator (2.101).

There are as expected four distinct regions. In the lowest region, $[b/r, -J'(u'(b))] = [14, 18.2815]$, there is consumption at the minimum possible level, and the investment in the risky asset gradually rises from 0. In the next region $[-J'(u'(b)), -J'(1)] = [18.2815, 23.9055]$, the wealth level is high enough to persuade the agent to risk some higher consumption. The next region is $[-J'(1), x^*] = [23.9055, 28.5487]$ where the agent consumes at the maximal level $\bar{c}$ but is not willing to raise that level. The final region lies to the right of x^*, where the agent raises the consumption level immediately to bring $x \equiv w/\bar{c}$ back down to x^*.

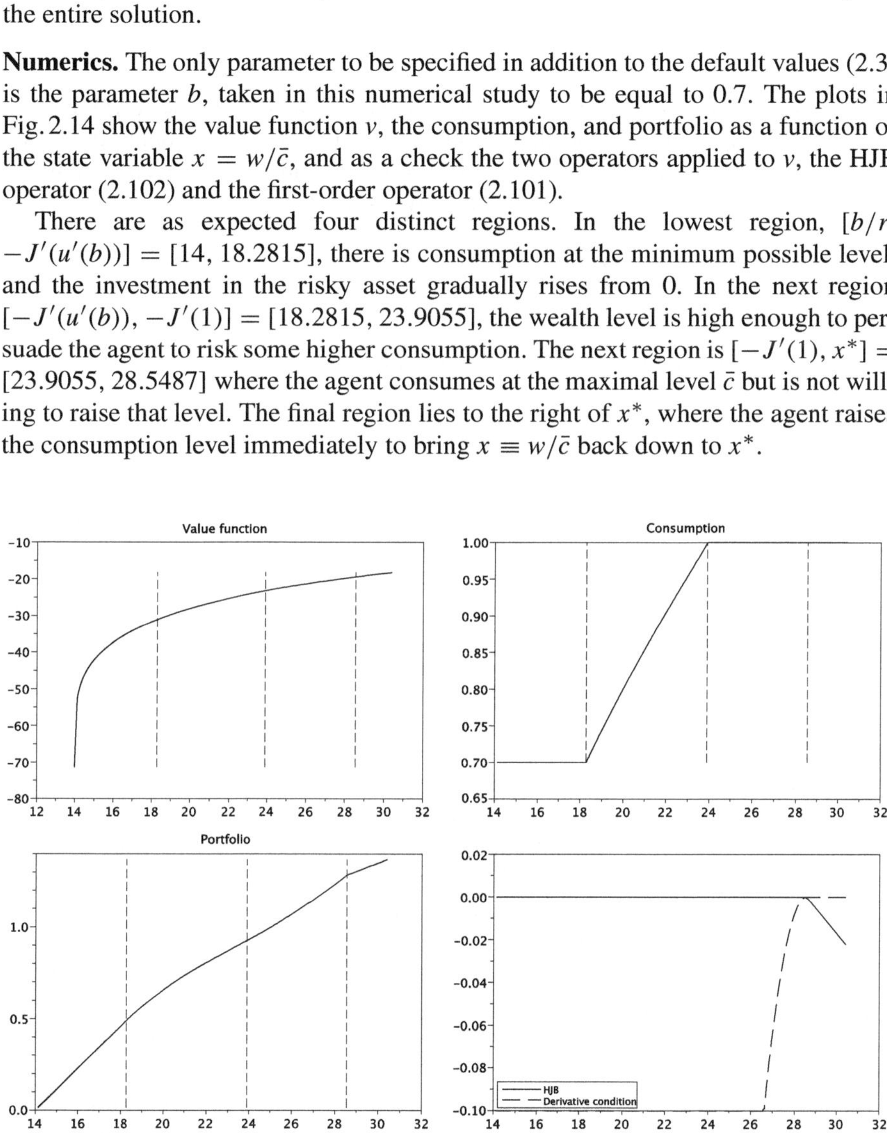

Fig. 2.14 Plots of the value, consumption, portfolio as a function of $x = w/\bar{c}$ for the problem with bounded drawdown of consumption, together with the check of the HJB variational inequality

The fourth plot shows that the maximum of the two is everywhere zero, as it should be, with the HJB holding with equality everywhere except the right-most region, where the linear operator applied to v is zero, again as expected.

2.16 Option to Stop Early

In this example, we once again assume the standard wealth dynamics (2.1), but we allow the possibility that the agent may choose to stop at some stopping time τ of his choice; when he stops he receives an immediate reward of $F(w_\tau)$, and that is the end of consumption. Thus the agent's objective is to obtain

$$V(w) = \sup_{c\geq 0,\theta,\tau} E\left[\int_0^\tau e^{-\rho t} u(c_t)\, dt + e^{-\rho\tau} F(w_\tau) \,\middle|\, w_0 = w\right]. \tag{2.107}$$

The Martingale Principle of Optimal Control tells us that

$$Y_t = V(w_t)e^{-\rho t} I_{\{t<\tau\}} + F(w_\tau)e^{-\rho\tau} I_{\{t\geq\tau\}} + \int_0^{t\wedge\tau} e^{-\rho s} u(c_s)\, ds$$

is a supermartingale and a martingale under optimal control. Using Itô's formula, we deduce that

$$0 \geq \sup\left[-\rho V + u(c) + (rw + \theta(\mu - r) - c)V' + \tfrac{1}{2}\sigma^2\theta^2 V''\right] \tag{2.108}$$

$$V \geq F, \tag{2.109}$$

with equality in at least one of these for each w. Even if we assume that u is CRRA, there is no scaling simplification possible because of the option to stop. However, we can still get a long way with this problem.

Firstly, notice that even though we have not assumed that F is concave, we may without loss of generality assume that it is, by replacing F by its least concave majorant $\bar{F}$. This is because if we were at some wealth level w where $F(w) < \bar{F}(w)$, we could turn up the value of θ to some vast number for a short time, until we reached one end or the other of the interval $[a, b]$ containing w in which $F < \bar{F}$. For vast values of θ, the volatility of the wealth process overwhelms the drift, so what we see is in effect a Brownian motion; accordingly, if we were at w we would have the option to stop at whichever of a or b the Brownian motion reached first, and the expected stopping value would just be the convex combination of $\bar{F}(a) = F(a)$ and $\bar{F}(b) = F(b)$, that is, $\bar{F}(w)$. So the agent wanting to stop at wealth level w could by this device improve his reward from $F(w)$ to $\bar{F}(w)$, and would of course do so.

We see from the HJB Equation (2.108) that the second derivative V'' must be everywhere non-positive, so we seek a concave function V dominating the concave function F. Taking dual variable $z \equiv V'$, and setting $J(z) = V(w) - zw$, we have

$J' = -x$, $J'' = -1/V''$, and the dual HJB equations become

$$0 \geq \tilde{u}(z) - \rho J(z) + (\rho - r)zJ'(z) + \tfrac{1}{2}\kappa^2 z^2 J''(z) \qquad (2.110)$$

$$J(z) \geq \tilde{F}(z). \qquad (2.111)$$

Example. Naturally, we have to be more explicit about the form of F and u in order to make more progress, so we shall assume that $u'(x) = x^{-R_1}$ and $F'(x) = x^{-R_2}$ for some $R_2 > R_1 > 1$. Since F converges to zero much faster than u as $x \to \infty$, we expect that the optimal rule will be to stop if and only if $w \geq w^*$ for some critical value w^* of w. In terms of the dual variable, this is equivalent to the statement that for $z \leq z^* = V'(w^*)$ we have equality in (2.111), else we have equality in (2.110). Hence we shall have for some constants A and B that

$$\begin{aligned} J(z) &= \tilde{F}(z) \qquad (z \leq z^*) \\ &= -\frac{\tilde{u}(z)}{Q(1 - R_1^{-1})} + A(z/z^*)^{-\alpha} + B(z/z^*)^{\beta} \qquad (z \geq z^*) \end{aligned}$$

where $-\alpha < 0 < 1 < \beta$ are the roots of the quadratic $Q(t) \equiv \frac{1}{2}\kappa^2 t(t-1) + (\rho - r)t - \rho$.

For large z, in order that J remains convex and decreasing, it has to be that $B = 0$ (since $\beta > 1$), so we just have to choose A and z^* to make J a C^1 function.

The equations determining A and z^* are (with $q = -Q(1 - R_1^{-1})$)

$$\begin{aligned} \tilde{F}(z^*) &= A + \tilde{u}(z^*)/q \\ (1 - R_2^{-1})\tilde{F}(z^*) &= -\alpha A + (1 - R_1^{-1})\tilde{u}(z^*)/q \end{aligned}$$

which gives

$$(\alpha + 1 - R_2^{-1})\tilde{F}(z^*) = (\alpha + 1 - R_1^{-1})\tilde{u}(z^*)/q$$

whence

$$(z^*)^{R_1^{-1} - R_2^{-1}} = \frac{\alpha + 1 - R_1^{-1}}{q(\alpha + 1 - R_2^{-1})}$$

and

$$z^* = \left\{ \frac{\alpha + 1 - R_1^{-1}}{q(\alpha + 1 - R_2^{-1})} \right\}^{R_1 R_2/(R_2 - R_1)}.$$

This is a pleasingly explicit solution, though without some special features as in this example we will be forced to seek a numerical solution.

2.17 Optimization under Expected Shortfall Constraint

In this example, we suppose the standard asset dynamics (2.1), but with zero consumption: the goal here is to maximize a terminal wealth objective

$$\sup_\theta Eu(w_T) \quad \text{subject to} \quad E[(\bar{w}-w_T)^+] \le \alpha \tag{2.112}$$

for some constants α and $\bar{w}$. As we have seen in Section 1.4, we may choose any terminal wealth w_T subject to the budget constraint

$$E[\zeta_T w_T] = w_0,$$

so we now have an optimization problem constrained by two scalar constraints. We may rewrite the expected-shortfall constraint in terms of the function $g(w) \equiv \min\{0, w-\bar{w}\}$ to read

$$E[g(w_T)] + \alpha = z \ge 0$$

for the non-negative slack variable z. At this stage we translate the problem into Lagrangian form with multipliers λ, η to become

$$L(\lambda,\eta) = \sup_{w_T, z\ge 0} E\big[u(w_T) + \lambda(w_0 - \zeta_T w_T) + \eta\{\alpha + g(w_T) - z\}\big],$$

and observe that non-negativity of z forces $\eta \ge 0$ for dual feasibility, and $\eta z = 0$. Therefore the optimization problem in Lagrangian form is simply

$$L(\lambda,\eta) = \sup_{w_T\ge 0} E\big[u(w_T) + \lambda(w_0 - \zeta_T w_T) + \eta\{\alpha + g(w_T)\}\big], \tag{2.113}$$

where η is understood to be non-negative. The function $f(w) \equiv u(w) + \eta g(w)$ is therefore concave increasing, and the optimization of the Lagrangian form is achieved when

$$f'(w_T) = \lambda\zeta_T.$$

Substituting this into the Lagrangian form, the maximized value is

$$\begin{aligned} L(\lambda,\eta) = {} & E\big[\tilde{u}(\lambda\zeta_T) : \lambda\zeta_T < u'(\bar{w})\big] + E\big[u(\bar{w}) - \lambda\zeta_T\bar{w} : u'(\bar{w}) < \lambda\zeta_T < \eta + u'(\bar{w})\big] \\ & + E\big[\tilde{u}(\lambda\zeta_T - \eta) - \eta\bar{w} : \lambda\zeta_T > u'(\bar{w}) + \eta\big] + \lambda w_0 + \eta\alpha. \end{aligned} \tag{2.114}$$

Now we have that $\zeta_t = \exp(-\kappa W_t - (r + \frac{1}{2}\kappa^2)t)$, so the first two of the expectations appearing in (2.114) can be evaluated explicitly in terms of the cumulative Gaussian distribution function. In more detail,

$$\lambda\zeta_T < q \Leftrightarrow W_T > \{\log\lambda - (r + \tfrac{1}{2}\kappa^2)T - \log q\}/\kappa \equiv \psi(q).$$

Writing $b = \psi(u'(\bar{w}))$ and $a = \psi(\eta + u'(\bar{w}))$, and assuming that $u'(x) = x^{-R}$ for some $R \neq 1$ we find that the first expectation in (2.114) is expressed as

$$\tilde{u}(\lambda) E\left[(\zeta_T)^{1-1/R} : W_T > b\right],$$

the second as

$$u(\bar{w}) P[a < W_T < b] - \lambda \bar{w} E[\zeta_T; a < W_T < b],$$

and the third as

$$E[\tilde{u}(\lambda\zeta_T - \eta) : W_T < a] - \eta \bar{w} P[W_T < a].$$

With the exception of the expectation in the last of these terms, everything can be evaluated explicitly in terms of the standard Gaussian distribution function, and even this expectation can be rapidly evaluated as it is a one-dimensional integral of a well-behaved function.

Numerics. For a numerical example, take $w_0 = 1$, $T = 10$, and $\alpha = 0.01$, with the other parameters as usual (2.3). Suppose that the shortfall value $\bar{w}$ to be compared with is the value that the initial cash would have achieved if invested solely in the bank account, for this example, $\bar{w} = 1.6487$. The unconstrained Merton investor will finish with terminal wealth equal to $I(\lambda'\zeta_T)$ for some λ' which matches the initial wealth condition,[14] whereas the terminal wealth of the shortfall-constrained investor will be a different function of the state-price density at time T. Figure 2.15 shows the wealth achieved by the constrained investor as a function of the wealth of the unconstrained investor, with the diagonal shown as a dashed line. The efficiency of the constrained investor in this example is 92.75 %.

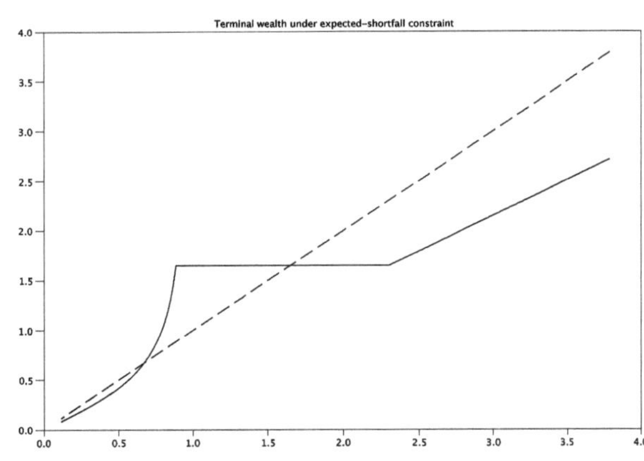

Fig. 2.15 Plot of the wealth achieved by the constrained investor as a function of the wealth achieved by the unconstrained investor

[14] In fact, we have $\lambda' = \exp(-\gamma_0 RT) w_0^{-R}$, where $\gamma_0 = (R-1)(r + \kappa^2/2R)/R$.

The qualitative features are very natural; for very high wealth, the unconstrained investor is getting more, but for wealths around the comparison value $\bar{w}$ the constrained investor receives just the riskless return (in this example, the risk-neutral probability that the constrained agent receives only the risk-neutral return is 56.68 %, and the time-0 cost of funding this certain payout in the event that it should be required is 0.4576). Once the wealth levels get very low, the wealth of the constrained agent falls below the wealth of the unconstrained agent, though this is somehow unimportant since these outcomes are very unlikely.

2.18 Recursive Utility

This example takes the usual wealth dynamics (2.1) but now has the unconventional recursive utility objective of maximizing U_0, where $(U_t)_{0\le t\le T}$ is a recursive utility process satisfying

$$Y_t \equiv U_t + \int_0^t F(s, c_s, U_s)\, dt = E\left[\int_0^T F(s, c_s, U_s)\, ds + G(w_T)\,\middle|\,\mathscr{F}_t\right] \quad (2.115)$$

where we suppose that F is concave increasing in its last two arguments, and that G is concave increasing. In general, it is not obvious that there should be *any* process U to solve the Eq. (2.115); any such process U solves an SDE, but with a *terminal* condition $U_T = G(w_T)$. General results on the existence and uniqueness of such backward SDEs (BSDEs) are well known, however; see [12] for an excellent survey of various applications in finance. In the simple setting of time-invariant dynamics, we expect that it will be possible to express $U_t = V(t, w_t)$ for some function V which we need to find. If this is the case, then the MPOC would lead us to expect that the process Y will be a supermartingale under any control, and a martingale under optimal control. This gives us the HJB equation

$$0 = \sup_{c\ge 0,\theta}\left[V_t + (rw + \theta(\mu - r) - c)V_w + \tfrac{1}{2}\theta^2\sigma^2 V_{ww} + F(t, c, V)\right]. \quad (2.116)$$

To illustrate how this would work, we shall take an example where

$$F(t, c, V) = e^{-\rho t} c^{\alpha} V^{\beta} \quad (2.117)$$

for some constants $\rho > 0, \alpha,\ \beta \in (0, 1)$. We conjecture that $V(t, w) = e^{-\nu t}\varphi(w)$ for some constant ν, which surprisingly turns out *not* to be the discount rate ρ appearing in the definition of F. Indeed, if we substitute the conjectured form of V into (2.116), this becomes

$$0 = \sup_{c\geq 0,\theta} e^{-\nu t}\left[-\nu\varphi + (rw + \theta(\mu - r) - c)\varphi' + \tfrac{1}{2}\theta^2\sigma^2\varphi'' + e^{-\rho t}c^{\alpha}e^{\nu(1-\beta)t}\varphi^{\beta}\right], \tag{2.118}$$

which leads to a time-invariant solution only if

$$\nu = \rho/(1-\beta). \tag{2.119}$$

Assuming this, optimizing over c gives the optimal choice:

$$\alpha c^{\alpha-1} = \varphi'/\varphi^{\beta}, \tag{2.120}$$

and optimizing the quadratic gives finally

$$0 = -\nu + rw\varphi' - \tfrac{1}{2}(\kappa\varphi')^2/\varphi'' + (1-\alpha)(\alpha\varphi')^{\alpha/(\alpha-1)}\varphi^{\beta/(1-\alpha)}. \tag{2.121}$$

This non-linear ODE is not soluble in closed form, but we can use dual variables to transform the problem to the more tractable form

$$0 = -\nu J + (\nu - r)zJ' + \tfrac{1}{2}\kappa^2 z^2 J'' + (1-\alpha)(\alpha z)^{\alpha/(\alpha-1)}(J - zJ')^{\beta/(1-\alpha)}. \tag{2.122}$$

Numerics. Figure 2.16 presents plots of the value function φ, the optimal portfolio divided by wealth, and the optimal consumption rate divided by wealth. The parameters used for the plots are $\alpha = 0.5$, $\beta = 0.4$. Notice how the middle (portfolio) plot falls with wealth, while the lower (consumption) plot rises with wealth, and contrast this with Fig. 2.7. In this situation, raising consumption is doubly important, not just because the running integral contribution to the objective rises directly with consumption, but also because it rises *indirectly* with consumption through the effect of higher U_t.

2.19 Keeping up with the Jones's

This is an example with two agents each playing the standard wealth dynamics (2.1), but where the utility of each agent depends on how much the other is consuming: the objective of agent i this time is to obtain

$$\sup E\int_0^\infty U_i(c_i(t), c_{1-i}(t))\,dt \qquad (i = 0, 1). \tag{2.123}$$

We can treat this by the static programming approach explained in Section 1.4. If the other agent has fixed his choice of consumption stream, then we must have

$$U_i'(c_i(t), c_{1-i}(t)) = \lambda_i\zeta_t \tag{2.124}$$

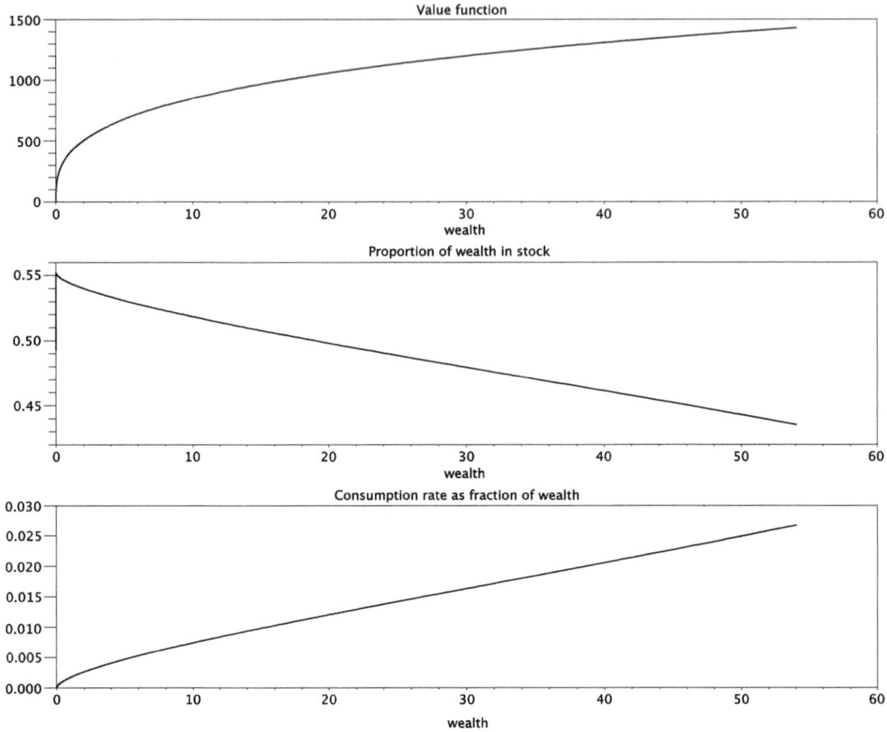

Fig. 2.16 Plots of the value, portfolio and consumption rates for the recursive utility example of Section 2.18

for some scalars λ_0, λ_1 chosen to satisfy the budget constraints. This gives us two equations for two unknowns which should in principle be soluble. To help us make progress, we shall suppose the simple form

$$U_i(c_i, c_{1-i}) = \frac{c_i^{1-R_i}}{1-R_i} \left(\frac{c_{1-i}}{c_i} \right)^{\alpha_i}, \tag{2.125}$$

where we will assume that $R_i > 1$, and $\alpha_i > 0$ so as to guarantee the property that as the other agent consumes more, you feel less happy, but you are always happier when you consume more. Some straightforward calculations now lead to the conclusion that

$$c_i \propto \zeta^{-\beta_i} \tag{2.126}$$

where

$$\beta_i \equiv \frac{\alpha_0 + \alpha_1 + R_{1-i}}{R_0 R_1 + \alpha_1 R_0 + \alpha_0 R_1}. \tag{2.127}$$

Structurally this looks like each agent behaves like a standard Merton investor with coefficient of relative risk aversion equal to

$$\tilde{R}_i = \frac{R_0 R_1 + \alpha_1 R_0 + \alpha_0 R_1}{\alpha_0 + \alpha_1 + R_{1-i}} = R_i + \frac{\alpha_i (R_{1-i} - R_i)}{\alpha_0 + \alpha_1 + R_{1-i}}. \tag{2.128}$$

Thus each agent's effective coefficient of relative risk aversion gets moved towards the other's; the more risk averse becomes less risk averse, and *vice versa*.

2.20 Performance Relative to a Benchmark

Frequently a fund manager will be judged by his ability to beat a benchmark. Thus if the benchmark process is the positive semimartingale q, the objective of the fund manager is

$$\sup E\, u(w_T/q_T) \tag{2.129}$$

where $T > 0$ is some fixed time horizon, and u is a given utility. Performance relative to a benchmark is really only an interesting question if there are many assets to invest in, so we shall assume the standard complete multivariate market (1.10). At one level, the solution is very easy. Using the static programming approach, Section 1.4, we see that we may achieve any terminal wealth w_T subject to the budget constraint

$$E[\zeta_T w_T] \le \zeta_0 w_0,$$

so we simply absorb this constraint with a Lagrange multiplier, and solve the unconstrained problem

$$\sup E\left[u(w_T/q_T) + \lambda(\zeta_0 w_0 - \zeta_T w_T)\right] \tag{2.130}$$

and then directly optimizing we obtain that

$$u'(w_T/q_T) = \lambda \zeta_T q_T, \tag{2.131}$$

which characterizes the optimal terminal wealth up to a relatively unimportant scalar multiple. Thus the optimal wealth process is represented as

$$\zeta_t w_t = E_t\left[\zeta_T q_T I(\lambda \zeta_T q_T)\right], \tag{2.132}$$

which in the case of a CRRA utility u becomes simply

$$\zeta_t w_t = \lambda^{-1/R} E_t\left[(\zeta_T q_T)^{1-1/R}\right]. \tag{2.133}$$

The extent to which we can solve this problem explicitly depends on the extent to which we can represent the martingale (2.133).

Note that most market indices, such as the FTSE100, the DJIA, the S&P500 are arithmetic averages of the individual component prices; the FT30 however is a geometric average, so the mathematically tractable idealization of a geometric average does exist, even if it is a bit unusual.

2.21 Utility from Slice of the Cake

Here is an example where an agent's preferences over consumption streams depend on what is happening to others, as in the example of keeping up with the Jones's, Section 2.19.

A continuous-time model of an economy contains a single productive asset, whose output process $(\delta_t)_{t\geq 0}$ evolves as

$$d\delta_t = \delta_t(\sigma dW_t + \mu dt), \tag{2.134}$$

where W is a standard Brownian motion. Agent $i \in \{1, \ldots, J\}$ has preferences over consumption streams $(c_t^i)_{t\geq 0}$ given by

$$E\int_0^\infty e^{-\rho_i t} u_i(p_t^i)\, dt, \tag{2.135}$$

where

$$p_t^i = \frac{c_t^i}{\sum_j c_t^j} \tag{2.136}$$

and $u_i : (0, \infty) \to \mathbb{R}$ is C^2, strictly increasing and strictly concave, $u_i'(0) = \infty$, $u_i'(\infty) = 0$. Agent i initially holds a fraction π_0^i of the productive asset; what is the equilibrium allocation of the output of the economy?

In equilibrium, there are no mutually beneficial trades remaining between the agents. So let's consider a deal to be entered into at time s to receive an infinitesimal quantity of consumption Y at later time t. The marginal price $\Pi_{st}^i(Y)$ which agent i would be prepared to pay for this would satisfy

$$\Pi_{st}(Y)\, e^{-\rho_i s} u_i'(p_s^i)(1 - p_s^i)/C_s = E_s[Y\, e^{-\rho_i t} u_i'(p_t^i)(1 - p_t^i)/C_t] \tag{2.137}$$

where $C_t = \sum_j c_t^j$, since increasing c^i by infinitesimal ε increases p^i by infinitesimal $\varepsilon(1 - p^i)/C$. Thus agent i's state-price density process is of the form

$$\zeta_t^i = \frac{e^{-\rho_i t} u_i'(p_t^i)(1 - p_t^i)}{C_t}. \tag{2.138}$$

Since the filtration is that of a univariate Brownian motion, the market is complete, and therefore all agents have the same state-price density process (up to a scalar multiple). Hence for some $\lambda_i > 0$,

$$\frac{e^{-\rho_i t} u_i'(p_t^i)(1 - p_t^i)}{C_t} = \lambda_i \zeta_t, \tag{2.139}$$

where ζ is the common state-price density.

Noticing that $x \mapsto g_i(x) \equiv u_i'(x)(1-x)$ is decreasing from ∞ to 0 on (0, 1), we may re-express this as

$$g_i(p_t^i) = \lambda_i e^{\rho_i t} \zeta_t C_t,$$

so if h_i is inverse to g_i we learn that

$$p_t^i = h_i(\lambda_i e^{\rho_i t} \zeta_t C_t). \tag{2.140}$$

Summing on i gives the market-clearing condition

$$1 = \sum_i h_i(\lambda_i e^{\rho_i t} \zeta_t \delta_t), \tag{2.141}$$

since the total consumption must match the total output. One consequence of this is that for a given set of λ_i, for each t the product $\zeta_t \delta_t$ is deterministic. Thus p_t^i is a function only of t, since when markets clear we have $C_t = \delta_t$. The equilibrium price of the asset is given by

$$S_t = \zeta_t^{-1} E_t\Big[\int_t^\infty \zeta_s \delta_s \, ds\Big] \tag{2.142}$$

$$= \varphi(t)\delta_t \tag{2.143}$$

for some deterministic function φ, but this is about as far as we can get in general.

Notice that if $\rho_i = \rho$ for all i, then from (2.141) it must be that $e^{\rho t}\zeta_t \delta_t$ is constant. Looking at (2.140), we conclude that the fraction of the cake being consumed by agent i never changes!

2.22 Investment Penalized by Riskiness

Suppose we have a standard complete multi-asset log-Brownian market (1.10):

$$dS_t^i / S_t^i = \sum_{j=1}^d \sigma_{ij} dW_t^j + \mu_i dt,$$

and you appoint a manager to invest your initial wealth w_0 up until time T. If he chooses portfolio proportions π, then the wealth of the portfolio evolves as

$$dw_t/w_t = r dt + \pi_t(\sigma dW_t + (\mu - r)dt).$$

The manager claims to be able to detect trends in the asset prices, but you are sceptical; you do not know his secret methods, but you can certainly observe the volatility $|\sigma^T \pi_t|$ of the wealth process, and you agree to pay him at time T the amount

$$x_T \equiv a w_T \exp\big(-\tfrac{1}{2}\varepsilon \int_0^T |\sigma^T \pi_s|^2 \, ds\big),$$

where $a, \varepsilon > 0$. By penalizing him according to the realized volatility of his strategy, you hope to prevent him pursuing risky strategies at your expense. If the manager's objective is to maximize $Eu(y_T)$, where u is CRRA, $u'(x) = x^{-R}$, what will he do?

To see what happens, define

$$x_t \equiv w_t \exp\big(-\tfrac{1}{2}\varepsilon \int_0^t |\sigma^T \pi_s|^2 \, ds\big),$$

and let

$$V(t, x) = \sup E_t[u(x_T) \mid x_t = x]$$

be the value function for the manager. The evolution of x is given by

$$dx_t = x_t \big[r dt + \pi_t \cdot (\sigma dW_t + (\mu - r)dt) - \tfrac{1}{2}\varepsilon |\sigma^T \pi_t|^2 dt \big],$$

and hence from the Martingale Principle of Optimal Control, we deduce the HJB equations

$$0 = \sup_{\pi} \big[V_t + x(r + \pi \cdot (\mu - r) - \tfrac{1}{2}\varepsilon |\sigma^T \pi|^2) V_x + \tfrac{1}{2} |\sigma^T \pi|^2 x^2 V_{xx} \big].$$

Now by scaling, we expect that $V(t, x) = f(t)u(x)$ for some function f of time, and substituting this form into the HJB equations we learn that

$$0 = \sup_{\pi} \; u(x) \big\{ \dot{f} + (1 - R) f (r + \pi \cdot (\mu - r) - \tfrac{1}{2}\varepsilon |\sigma^T \pi|^2) - \tfrac{1}{2}(1 - R) R \, f \, |\sigma^T \pi|^2 \big\}.$$

Thus the optimal portfolio choice for the manager is to use

$$\pi = (R + \varepsilon)^{-1} (\sigma\sigma^T)^{-1} (\mu - r),$$

which is the optimal portfolio choice for a Merton investor with constant coefficient of risk aversion $R + \varepsilon$; by introducing the penalty for portfolio volatility, *you have increased the manager's effective risk-aversion by ε!*

2.23 Lower Bound for Utility

This example[15] assumes the standard wealth dynamics (2.1) with running consumption, but now we suppose that the utility of the agent is bounded below, but his wealth is not. The basic example concludes that as an agent's wealth falls lower and lower, so does his consumption; but this does seems to be counter to human behaviour. If an individual's wealth is so low that he would be reduced to starving gradually to death, we do not expect him meekly to accept his demise; in reality, he would beg, borrow or steal the wherewithal of living. The worst that could happen to him would be that he gets found out and thrown into jail, and that would be the same outcome whether he had borrowed \$2000 or \$2M. So we will suppose that the agent may borrow or steal money to support a higher-than-starvation level of consumption; in other words, we relax the constraint that wealth should be non-negative.

Once we do this, there have to be other modifications to the problem specification to prevent it becoming trivial. If he is allowed to go into negative wealth, why does he not just borrow indefinitely and enjoy himself with other people's money? So we introduce the possibility of his finances being reviewed, according to a variable intensity

$$G(w,\theta^2) = (b|w|^m + a\theta^2) I_{\{w<0\}}, \tag{2.144}$$

where $a,\ b,\ m$ are positive. If the agent gets reviewed while his wealth is negative, he is found out and thrown into jail, incurring a (large) negative penalty $-K$. Thus his objective is

$$V(w) = \sup_{\theta,c} E\left[\int_0^\tau e^{-\rho t} u(c_t)\,dt - e^{-\rho\tau} K \,\middle|\, w_0 = w\right], \tag{2.145}$$

where τ denotes the time of the first review.

A few comments on the modelling assumptions are needed here. Firstly, we assume the review intensity is zero while wealth is positive. This is not to say that an individual's affairs might not be scrutinized while his wealth is positive, but if they were, then he would be found to be living honestly and allowed to continue. So we lose nothing by ignoring such reviews. Next, the requirement that the review intensity depends on θ^2 corresponds to a plausible feature, that if the agent was investing enormous amounts in the risky asset he would attract the attention of regulators; mathematically, we need this feature, because otherwise the agent faced with negative wealth could turn up θ to some huge value, and then move rapidly through negative values of wealth until he got back to positive wealth again. For very large θ, the volatility of the wealth overwhelms any drift effect, so we are seeing wealth evolve in effect as a very fast Brownian motion; what we have is a doubling strategy. To rule this out, we suppose that large risky positions greatly increase the chances of detection. The final observation is that unless the risk of detection got higher the

[15] An extended account can be found in Muraviev & Rogers [29].

more negative wealth becomes, then the impoverished agent could simply come out of the risky asset entirely, eliminate the risk of discovery, and borrow indefinitely to fund consumption.

At the random time τ of discovery, the agent's value falls from $V(w_{\tau-})$ to $-K$. Using the Martingale Principle of Optimal Control on the value function, we deduce the HJB equation for this problem:

$$0 = \sup_{c,\theta} \bigl[-\rho V + u(c) + (rw + \theta(\mu - r) - c)V' + \tfrac{1}{2}\sigma^2\theta^2 V'' - G(w, \theta^2)(K + V) \bigr]. \tag{2.146}$$

It is clear that the value for this problem must be always at least $-K$, so V cannot be globally concave. This alters the HJB equation somewhat, because when we optimize over θ, in places the term $\frac{1}{2}\sigma^2\theta^2 V''$ will be positive; the only thing that prevents the optimization over θ from becoming trivial is the presence of the final term $-G(w, \theta^2)(K + V)$, which is negative, and also quadratic in θ. This makes the problem more delicate numerically than many we have seen, and the route taken in [29] involves a variable transformation to restore concavity to the HJB equation; the interested reader is directed to [29] for all the details, but we will here just present some numerical results and leave it at that.

Numerics. It turns out to be notationally simpler to write $a = \sigma^2\nu^2/2$ in (2.144). In the example we present here, the values taken were $m = \nu = 2$, $K = 60$ and $b = 10$. We show the form of the value, the portfolio and the consumption rate for positive wealth values in Fig. 2.17. The dashed lines show the corresponding solution to the standard Merton problem, proportional to wealth in portfolio and consumption, as we know. Notice that as wealth increases, we see the solution approaching the Merton solution, not surprisingly; the very wealthy do not need to worry about bankruptcy!

We show in Fig. 2.18 the corresponding plots for negative wealth, and the first thing to notice is that the vertical scale is of a completely different order of magnitude from the plots for positive wealth; this was the reason for plotting them separately. The value falls gently to the asymptotic value $-K = -60$, but the portfolio rises dramatically; when wealth is -3.5, the cash value of the agent's holding of the risky asset is about 20, whereas you would need a positive wealth of about 50 to get such a large holding of the risky asset! What is happening is that the insolvent agent is gambling; the risky asset has a higher rate of return, so he is taking a chance that the higher growth rate will help him get back to positive wealth. At the same time, the rates of consumption are gigantic; in positive wealth, the consumption rate in the plot does not get above about 2, whereas in negative wealth the rate has exceeded 1000 when wealth has dropped to -2! You could interpret this as the agent turning to crime—he has abandoned hope of ever becoming honest again, and plunders as much as he can before being caught.

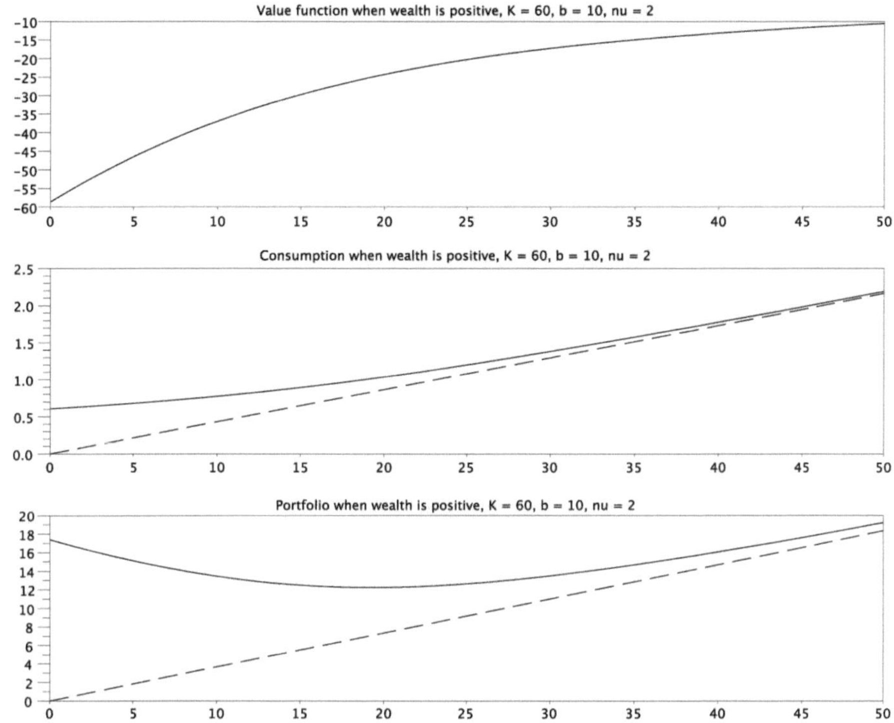

Fig. 2.17 Plots of the value, portfolio in the risky asset, and consumption rate for positive values of wealth, for the example of utility bounded below, Section 2.23

2.24 Production and Consumption

The story here is a little different; there is no financial market in which the agent is choosing to invest, but rather a real production process which generates an output. The agent's choice is how much of this output to consume.[16] We shall take the dynamical specification to be

$$dK_t = (I_t - \delta K_t)dt \tag{2.147}$$

$$Y_t = Z_t f(K_t) = I_t + C_t. \tag{2.148}$$

Here, K_t is the available capital at time t, which depreciates at rate δ and is replenished from output at rate I_t. The agent has to choose how to split the output Y_t between consumption C_t and investment. The output depends on K_t and on a random factor Z_t, where f is an increasing concave production function, and $dZ_t = Z_t(\sigma dW_t + \mu dt)$ is a log-Brownian motion. The agent has objective

[16] This is a very classical growth problem; see, for example, the book by Romer [36] for more background. We take here what is perhaps the simplest form of the problem.

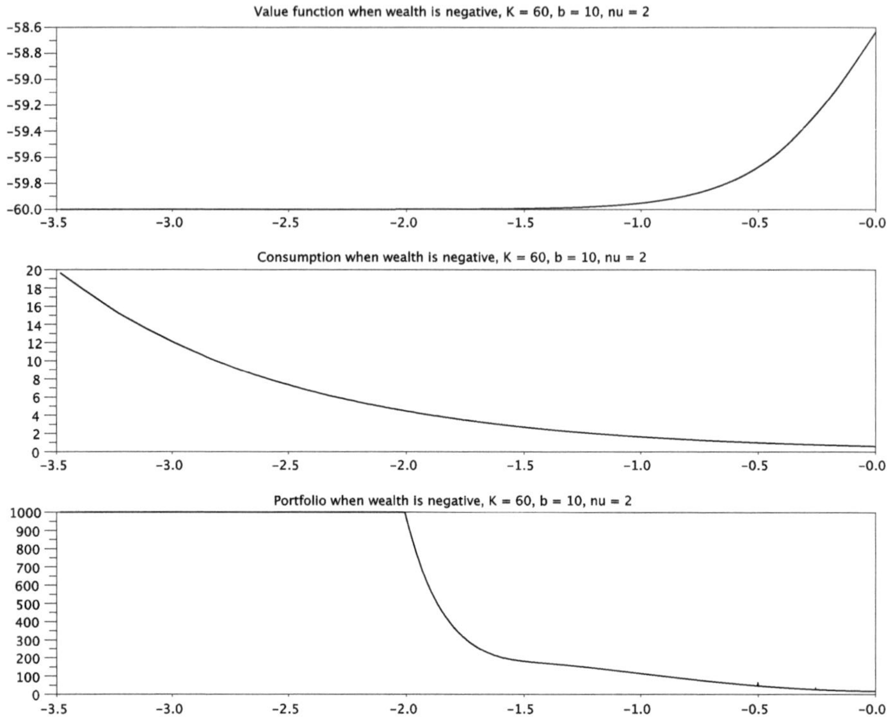

Fig. 2.18 Plots of the value, portfolio in the risky asset, and consumption rate for negative values of wealth, for the example of utility bounded below, Section 2.23. The consumption plot is truncated at 1000

$$V(z,k) \equiv \sup E\left[\int_0^\infty e^{-\rho t} u(C_t)\,dt \,\middle|\, Z_0 = z,\, K_0 = k\right]; \tag{2.149}$$

as usual, and there is a conflict between consuming now, and investing more to generate more output (and potentially more consumption) at later time.

To solve this, we can write down the HJB equation for the problem

$$0 = \sup_C \left[-\rho V + u(C) + \mu z V_z + \tfrac{1}{2}\sigma^2 z^2 V_{zz} + (zf(k) - \delta k - C)V_k\right], \tag{2.150}$$

where as usual we will assume that u is constant relative risk aversion, $u'(x) = x^{-R}$ for some $R > 1$ (the problem is ill posed if $0 < R < 1$). At this stage, we usually look for scaling properties to allow us to reduce the number of independent variables in the equation; but things are not so simple this time. Notice that if we were to double Z, we could look at (2.148) and think that we could then double I and C; but it's not that simple, because if we doubled investment the path of K would have changed. However, if we assume that

$$f(K) = AK^\alpha \tag{2.151}$$

for some $A > 0$ and $0 < \alpha \leq 1$, then if we take the time-0 state (z, k) and rescale to $(\lambda^{1-\alpha}z, \lambda k)$ for some $\lambda > 0$, then we have scaled I and C by λ, and therefore have scaled the objective by λ^{1-R}, that is,

$$V(\lambda^{1-\alpha}z, \lambda k) = \lambda^{1-R} V(z, k)$$

from which we conclude that

$$V(z, k) = k^{1-R}\, V(k^{\alpha-1}z, 1) \equiv u(k)\, h(x), \tag{2.152}$$

where we have taken $x \equiv k^{\alpha-1}z$.

Before we develop the HJB equation further, let us notice that the problem as originally posed can be reduced to the situation where $\delta = 0$, by setting

$$\tilde{K}_t = e^{\delta t} K_t, \quad \tilde{I}_t = e^{\delta t} I_t, \quad \tilde{C}_t = e^{\delta t} C_t, \quad \tilde{Z}_t = e^{(1-\alpha)\delta t} Z_t$$

so that the dynamics read

$$d\tilde{K}_t = \tilde{I}_t dt, \qquad e^{\delta t} Y_t = \tilde{Z}_t f(\tilde{K}_t) = \tilde{I}_t + \tilde{C}_t,$$

and the objective has become

$$E \int_0^\infty e^{-\rho t} e^{\delta(R-1)t}\, u(\tilde{C}_t)\, dt.$$

By changing ρ to $\rho' \equiv \rho - \delta(R-1)$ we reduce the original problem to the case where $\delta = 0$, but we also learn that we need the condition

$$\rho - \delta(R-1) > 0. \tag{2.153}$$

This condition has the following natural interpretation. Suppose that Z had dropped to zero; then there would be no output, and the only utility we could derive would be from consuming the capital. We therefore need to solve the deterministic optimization problem

$$\sup \int_0^\infty e^{-\rho t} u(C_t)\, dt \qquad \text{subject to} \quad \int_0^\infty C_t\, dt = K_0. \tag{2.154}$$

This problem is only well-posed if condition (2.153) holds, and in this case the optimal choice of C is

$$C_t = K_0\, e^{-\rho t/R} \rho / R.$$

The value of the problem is then seen to be (finite if (2.153) holds and then equal to)

$$u(K_0)\left(\frac{R}{\rho}\right)^R = u(K_0)h(0), \tag{2.155}$$

which tells us the value of h at zero.

Now we resume the analysis of the HJB equation. Using the scaling relationship (2.152) the equation (2.150) takes the form (with $y = C/k$, and recalling that $\delta = 0$ now)

$$\begin{aligned} 0 &= \sup_y u(k)\left[-\rho h + y^{1-R} + \mu x h' + \tfrac{1}{2}\sigma^2 x^2 h'' + \{(1-R)h + (\alpha-1)xh'\}(Ax - y)\right] \\ &= u(k)\Bigg[-\rho h + \mu x h' + \tfrac{1}{2}\sigma^2 x^2 h'' - \{(R-1)h + (1-\alpha)xh'\}Ax \\ &\qquad + R\left(h + \frac{1-\alpha}{R-1}xh'\right)^{(R-1)/R}\Bigg]. \end{aligned} \tag{2.156}$$

The optimal choice of y is

$$y^* = \frac{C^*}{K} = \left(h + \frac{1-\alpha}{R-1}xh'\right)^{-1/R}. \tag{2.157}$$

Using the abbreviation $b = (R-1)/(1-\alpha)$, and expressing $h(x) = g(w)$ with $w \equiv \log x$ turns (2.156) into

$$0 = -\rho g + \mu g' + \tfrac{1}{2}\sigma^2 g'' - \{(R-1)g + (1-\alpha)g'\}Ae^w + R\left(g + \frac{1-\alpha}{R-1}g'\right)^{(R-1)/R} \tag{2.158}$$

Numerics. For the numerical example, we took $\alpha = 0.7$ and $A = 2$; the depreciation δ was supposed to be zero, as explained above. The first plot shows the value function h, and below it the consumption rate $y = C/K$.

Interestingly, the consumption rate appears almost proportional to Z; the expression (2.157) for y^* as a function of x is nearly linear in x. However, near to $x = 0$ the consumption does not fall away entirely to zero, because even if the random factor Z (and therefore output) is very small, the agent will still consume out of the capital (Fig. 2.19).

2.25 Preferences with Limited Look-Ahead

The standard objective (2.2) of an agent involves consideration of consumption at all future times, and its analysis is based on strong assumptions to be made on the dynamics of the processes for all time. In practice, such assumptions are hard to

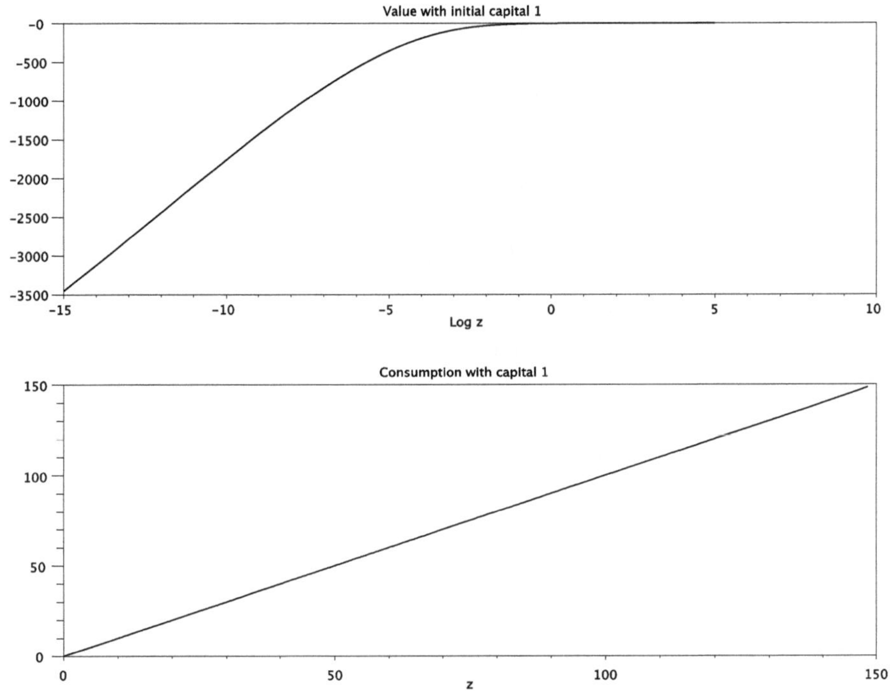

Fig. 2.19 Plots of the value and consumption rate for the example of Section 2.24 of an economy with production and consumption

defend, and the mental picture of an agent reflecting on the possible outcomes of his investments 60 years into the future is not one that most people would be familiar with.

So what would be a more plausible story? One answer would be one in which the agent cares about his consumption over the next T units of time, but thereafter he accepts that his uncertainty is so great that really all that he can say is that he would prefer to get through the next T units of time with more wealth rather than less. So we might propose that what the agent cares about is

$$E_t \left[\int_t^{t+T} u(s-t, c_s)\, ds + g(w_{t+T}) \right] \tag{2.159}$$

for some increasing concave g. We shall suppose that the agent takes the wealth dynamics (2.1) as given,[17] and aims to optimize his objective—but what does that mean? He might decide now at time t what his best actions would be, but at some later time $t+h < t+T$ he would have a different objective, and he might then want to (and would be able to) change what he had planned to do at time t. Such problems have been considered by Ekeland & Lazrak [11] and by Björk & Murgoci [3] who

[17] We even assume that the parameters are known.

formulate the problem as a game between the agent now and his later selves. The notion of solution is a Nash equilibrium; a choice of current actions which could not be improved if all of the later selves were to stick with their chosen actions.[18]

To explain this more concretely, suppose that the agent chooses to consume at rate $c_t = c(w_t)$ and invest $\theta_t = \theta(w_t)$ in the risky asset, for some suitable well-behaved functions c, θ. Then the controlled wealth process evolves as

$$dw_t = rw_t\,dt + \theta(w_t)(\sigma dW_t + (\mu - r)dt) - c(w_t)dt, \tag{2.160}$$

which is an autonomous diffusion. This being the case, we can in principle find the transition density of the diffusion, and could then calculate

$$\varphi(t, w) = E\left[\int_t^T u(s, c(w_s))\,ds + g(w_T)\,\middle|\, w_t = w\right],$$

which solves the Cauchy problem

$$\frac{\partial\varphi}{\partial t} + u(t, c(w)) + \mathscr{L}\varphi(t, w) = 0, \qquad \varphi(T, w) = g(w),$$

where $\mathscr{L}$ is the infinitesimal generator of the diffusion:

$$\mathscr{L} \equiv \tfrac{1}{2}\sigma^2\theta(x)^2\frac{\partial^2}{\partial x^2} + \{rx + \theta(x)(\mu - r) - c(x)\}\frac{\partial}{\partial x}.$$

The notion of solution is that φ should satisfy the HJB equations for the value *at time 0*:

$$\sup_{c,\theta}\left[\frac{\partial\varphi}{\partial t}(0, w) + u(0, c) + \tfrac{1}{2}\sigma^2\theta^2\frac{\partial^2\varphi}{\partial x^2}(0, w) + \{rw + \theta(\mu - r) - c\}\frac{\partial\varphi}{\partial x}(0, w)\right] = 0 \tag{2.161}$$

and that the supremum is attained by $c = c(w)$, $\theta = \theta(w)$.

In general it will be hard to make progress on this problem, but there is a simple example which can be worked through, and shows clearly the features of interest here. Let us suppose that $u(t, c) = h(t)u(c)$, $g(w) = Au(w)$, where $u'(c) = c^{-R}$ as in Section 2.1. The agent there (with a *fixed* time horizon T) will invest a fixed proportion π_M of his wealth in the risky asset at all times. However, he will in general *not* consume at a rate which is a constant multiple of his current wealth; see (2.9). In the present example where the agent has a fixed but rolling horizon, consumption is at a fixed multiple of wealth for all time; how do we decide what the agent does?

Suppose that the agent consumes at rate $c_t = aw_t$; then the wealth process is

$$w_t = w_0 \exp(\sigma\pi_M W_t + (b - a)t)$$

---

[18] In this case, because of the time-invariance of the problem, they would in fact be choosing the same actions as the current agent.

where

$$b \equiv r + \pi_M(\mu - r) - \tfrac{1}{2}\sigma^2\pi_M^2.$$

Routine calculations lead to the conclusion that

$$Eu(w_t) = u(w_0)\, e^{mt}$$

where

$$m = (R-1)(a - (r + \kappa^2/2R)). \tag{2.162}$$

Accordingly,

$$\varphi(t,w) = u(w)\left[\int_t^T h(s)\, e^{m(s-t)}\, a^{1-R}\, ds + Ae^{m(T-t)}\right],$$

and

$$\dot{\varphi}(t,w) + m\varphi(t,w) = -u(w)\, a^{1-R}\, h(t).$$

For brevity, we write

$$Q = \int_0^T h(s)\, e^{ms}\, a^{1-R}\, ds + Ae^{mT}, \tag{2.163}$$

so that the equation (2.161) to be solved becomes

$$\sup_{c,\theta}\left[-mQu(w) - a^{1-R}u(w)h(0) + (rw + \theta(\mu - r) - c)Qu'(w) - \tfrac{1}{2}\sigma^2\theta^2 Qu''(w) + h(0)u(c)\right] = 0. \tag{2.164}$$

Optimizing leads to the conclusion that

$$c = w(h(0)/Q)^{1/R}, \qquad \theta = \pi_M w. \tag{2.165}$$

Now Q depends in a reasonably complicated fashion on a, and for the choice a to constitute a Nash equilibrium choice we have to have

$$\frac{c}{w} = a = \left(\frac{h(0)}{Q}\right)^{1/R}, \tag{2.166}$$

which is an implicit equation to be solved for a. If a solves this equation, then it can be shown that (2.164) holds.

Let us see how this works out in the case where we take $h(t) = \exp(-\varepsilon t)$ for some $\varepsilon > 0$. In this case,

$$Q = \frac{1 - e^{-(\varepsilon - m)T}}{\varepsilon - m}\, a^{1-R} + Ae^{mT}$$

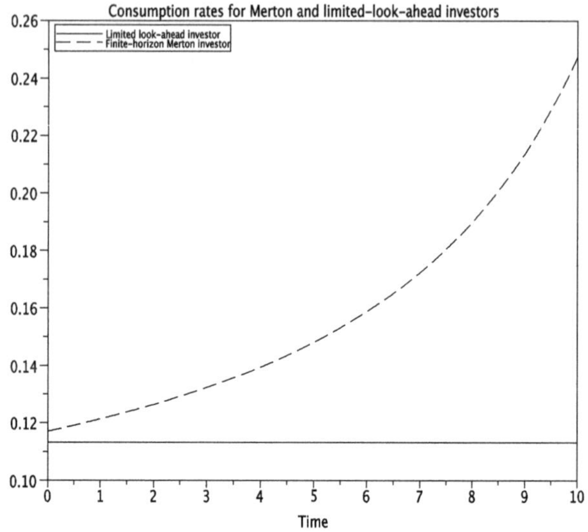

Fig. 2.20 Plots of the consumption rates for the investor with limited look-ahead and the corresponding Merton investor (Section 2.25)

where m depends on a as (2.162). This is to be compared with what happens when we do the usual finite-horizon optimization, as in Section 2.1. We saw there that the value function has the form $f(t)u(w)$ where f solves

$$\dot{f} - (R-1)(r + \kappa^2/2R)f + Rf^{1-1/R}h^{1/R} = 0, \qquad f(T) = A. \tag{2.167}$$

Numerics. In the numerical example we took $\varepsilon = 0.1$, $A = 6$, and $T = 10$. The results are shown in Fig. 2.20. As expected, the consumption rate of the investor with limited look-ahead remains constant, and below the consumption rate of the Merton investor. When the time horizon is still quite large, the two values are not far apart, 0.1132 compared with 0.1170. By the end of the time period, the Merton investor's consumption rate has risen to 0.2476. At the beginning of the time period, the difference in the solutions is relatively small, because the time horizon is 10 and the discounting rate ε is 0.1, so by the time horizon, the discounting has had quite an effect; nevertheless, the limited-lookahead investor is still being more cautious.

2.26 Investing in an Asset with Stochastic Volatility

In this section we will study a simple stochastic volatility model introduced in [20]. This model is an interesting stochastic volatility model because it gives rise to a complete market, so derivatives have unique prices.

The asset dynamics

$$dS_t = S_t(\sigma_t\, dW_t + \mu\, dt) \tag{2.168}$$

have stochastic volatility, but instead of supposing that this is driven by some independent process, we let the volatility be driven by the asset itself. In more detail, writing $X_t \equiv \log S_t$, we define the offset process Z by

$$Z_t = \int_{-\infty}^{t} \lambda e^{\lambda(s-t)}(X_s - X_t)\, ds \tag{2.169}$$

which measures how far the exponentially-weighted average of past log-price is above the current value. The stochastic volatility is now simply $\sigma_t = f(Z_t)$ for some function f to be specified. We have by some straightforward Itô calculus that

$$\begin{aligned} dZ_t + \lambda Z_t\, dt &= -dX_t \\ &= -\left\{ f(Z_t)\, dW_t + (\mu - \tfrac{1}{2} f(Z_t)^2)\, dt \right\} \end{aligned} \tag{2.170}$$

which exhibits Z as the solution[19] of an autonomous SDE, and therefore a diffusion.

The agent has the standard objective, so we have to identify the value function

$$V(w, z) = \sup E\left[\int_0^\infty e^{-\rho t} u(c_t)\, dt \,\middle|\, w_0 = w, z_0 = z \right]. \tag{2.171}$$

The value function solves the HJB equation

$$\begin{aligned} 0 = \sup_{c,\theta} \big[&-\rho V + u(c) + (rw + \theta(\mu - r) - c) V_w \\ &- (\lambda z + \mu - \tfrac{1}{2} f(z)^2) V_z + \tfrac{1}{2} f(z)^2 \left\{ \theta^2 V_{ww} - 2\theta V_{wz} + V_{zz} \right\} \big]. \end{aligned}$$

Assuming that u is CRRA as usual, scaling gives us the product form $V(w, z) = u(w)g(z)$ for the value, and the HJB equation now becomes (with $c = wx, \theta = wq$)

$$\begin{aligned} 0 = \sup_{x,q} u(w) \big[&-\rho g + x^{1-R} + (1-R)(r + q(\mu - r) - x)g - (\lambda z + \mu - \tfrac{1}{2} f^2) g' \\ &+ \tfrac{1}{2} f^2 \left\{ R(R-1) q^2 g - 2(1-R) q g' + g'' \right\} \big]. \end{aligned} \tag{2.172}$$

The optimizing choices are

$$x = g^{-1/R}, \qquad q = \frac{(\mu - r)g - f^2 g'}{R g f^2}, \tag{2.173}$$

and the HJB equation for g finally becomes

$$0 = -\rho g + R g^{1-1/R} + r(1-R)g - (\lambda z + \mu - \tfrac{1}{2} f^2) g' + \tfrac{1}{2} f^2 g'' + (1-R) \frac{((\mu - r)g - f^2 g')^2}{2 R g f^2}. \tag{2.174}$$

[19] Of course we need some conditions on f; bounded Lipschitz is quite sufficient.

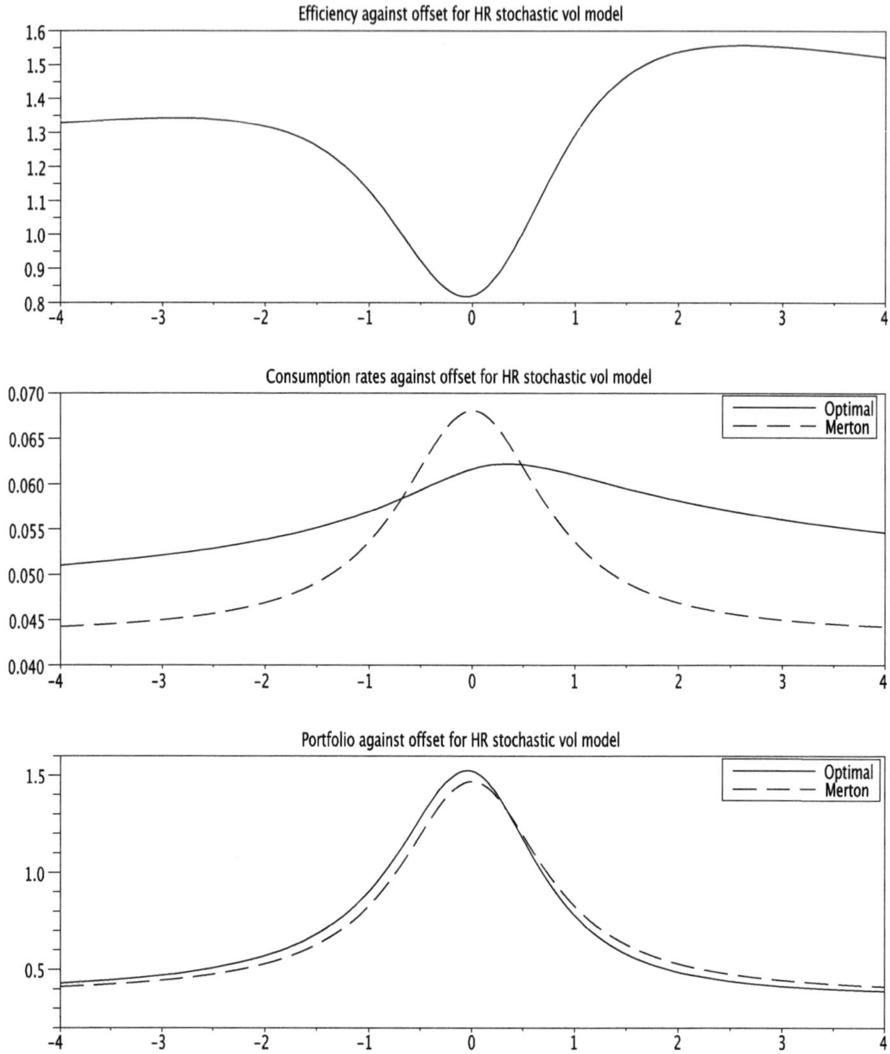

Fig. 2.21 Plots of efficiency, c/w and θ/w for the Hobson-Rogers stochastic volatility model of Section 2.26

Numerics. Figure 2.21 shows the results of a numerical study taking

$$f(z) = \frac{\sigma(1+z^2)}{2+z^2}, \qquad \lambda = 0.1. \tag{2.175}$$

The plots in Fig. 2.21 show the features of the solution as it depends on the offset z. In the top plot, we see the efficiency. To understand what this shows, when we compare the value $V(w,z) = u(w)g(z)$ with the value for the Merton problem, we need to specify *which* Merton problem; the natural thing to do is for each z to compare for the Merton problem where the volatility is constant and equal to $f(z)$. We note that the Merton value $\gamma_M^{-R} u(w)$ is decreasing with σ, all else being kept constant. So when z is near zero and volatility is at its lowest value, the stochastic volatility alternative should be worse, since the volatility can only get bigger if it changes. Far away from zero, we argue the other way round; the Merton situation with fixed high volatility is undesirable, but the stochastic volatility example has the chance to move back to lower volatility, so can be expected to do better. The asymmetry of the plot is explained by the fact that the SDE for the offset Z is not symmetric.

For consumption, the Merton investor with lower volatility will consume at a faster rate,[20] and a similar unimodal shape for the optimal consumption is visible, though smeared out as one would expect due to the variability of the volatility. Similar considerations apply for the portfolio proportions, though these are in fact remarkably close.

2.27 Varying Growth Rate

This is a story of a Bayesian agent, as in Section 2.32, but in this situation we do not suppose that the growth rate of the single risky asset is a constant, rather that it is evolving as a Brownian motion of small variance. This completely changes the form of the solution and the methods used to study it.

In this story, the risky asset dynamics are

$$dS_t = S_t(\sigma\, dW_t + \mu_t\, dt), \tag{2.176}$$

where σ is constant, but μ_t, the growth rate process, varies with time and has to be filtered from the observed prices. We denote by

$$Y_t \equiv \sigma^{-1} \log S_t \tag{2.177}$$

the observation process with dynamics

$$dY_t = dW_t + \alpha_t\, dt = dW_t + (\mu_t - \tfrac{1}{2}\sigma^2)\, dt/\sigma \tag{2.178}$$

and we propose that α is itself a Brownian motion with volatility ε:

$$d\alpha_t = \varepsilon\, dW_t' \tag{2.179}$$

where W' is a Brownian motion independent of W. The observation process Y generates a filtration $\mathcal{Y}_t \equiv \sigma(Y_s : s \leq t)$.

---

[20] Recall that $R = 2 > 1$.

We are now in the setting of the Kalman-Bucy filter (see, for example, [34], VI.9), which for simplicity we shall assume is in steady state. Defining the innovations Brownian motion ν by

$$dY_t = d\nu_t + \hat{\alpha}_t \, dt \tag{2.180}$$

where $\hat{\alpha}$ is the $\mathscr{Y}$-optional projection of α, it can be shown (see [34], VI.9) that

$$d\hat{\alpha}_t = \varepsilon \, d\nu_t. \tag{2.181}$$

The pair of Eqs. (2.180) and (2.181) are a compact representation of the asset dynamics. Now suppose that the agent has the standard objective (2.2) to optimize:

$$V(w,a) = \sup E\left[\int_0^\infty e^{-\rho t} u(c_t) \, dt \,\middle|\, w_0 = 0, \hat{\alpha}_0 = a \right]. \tag{2.182}$$

Assuming as so often that u is CRRA ($u'(x) = x^{-R}$) leads to the scaling relationship $V(w,a) = u(w) f(a)$ for some function f to be found. The HJB equation for this problem is

$$0 = \sup_{c,\theta} \left[-\rho V + u(c) + \left\{ rw + \theta(\sigma a + \tfrac{1}{2}\sigma^2 - r) - c \right\} V_w + \tfrac{1}{2}\sigma^2\theta^2 V_{ww} + \theta\varepsilon\sigma V_{wa} + \tfrac{1}{2}\varepsilon^2 V_{aa} \right].$$

Utilizing the scaling form of the solution, writing $x = c/w$, $q = \theta/w$, we find the HJB equation becomes

$$\begin{aligned} 0 = \sup_{x,q} u(w) \big[& -\rho f + x^{1-R} + (1-R)\left\{ r + q(\sigma a + \tfrac{1}{2}\sigma^2 - r) - x \right\} f \\ & - \tfrac{1}{2} R(1-R)\sigma^2 q^2 f + \tfrac{1}{2}\varepsilon^2 f'' + (1-R)\sigma\varepsilon q f' \big]. \end{aligned}$$

Calculus gives the optimality conditions

$$x^{-R} = f, \qquad \sigma^2 R q = \sigma a + \tfrac{1}{2}\sigma^2 - r + \sigma\varepsilon f'/f, \tag{2.183}$$

which turns the HJB equation into

$$0 = -\rho f + R f^{1-1/R} + r(1-R) f + \tfrac{1}{2}\varepsilon^2 f'' + \frac{(1-R) f}{2\sigma^2 R} (\sigma a + \tfrac{1}{2}\sigma^2 - r + \sigma\varepsilon f'/f)^2. \tag{2.184}$$

Numerics. The plots in Fig. 2.22 show how efficiency, consumption rate and portfolio vary with the posterior mean for the example where $\varepsilon = 0.2$. The efficiency is calculated by comparison with a standard Merton problem where the true mean is constant and equal to the posterior mean. We see that for $\hat{\alpha}$ near to zero the efficiency is high, then it drops away, then rises again. We can understand the peak at 0 by noting that if the mean is constant and equal to zero, then the stock is a bad investment, giving risk but no return; but if the posterior mean is zero, then there is the likelihood

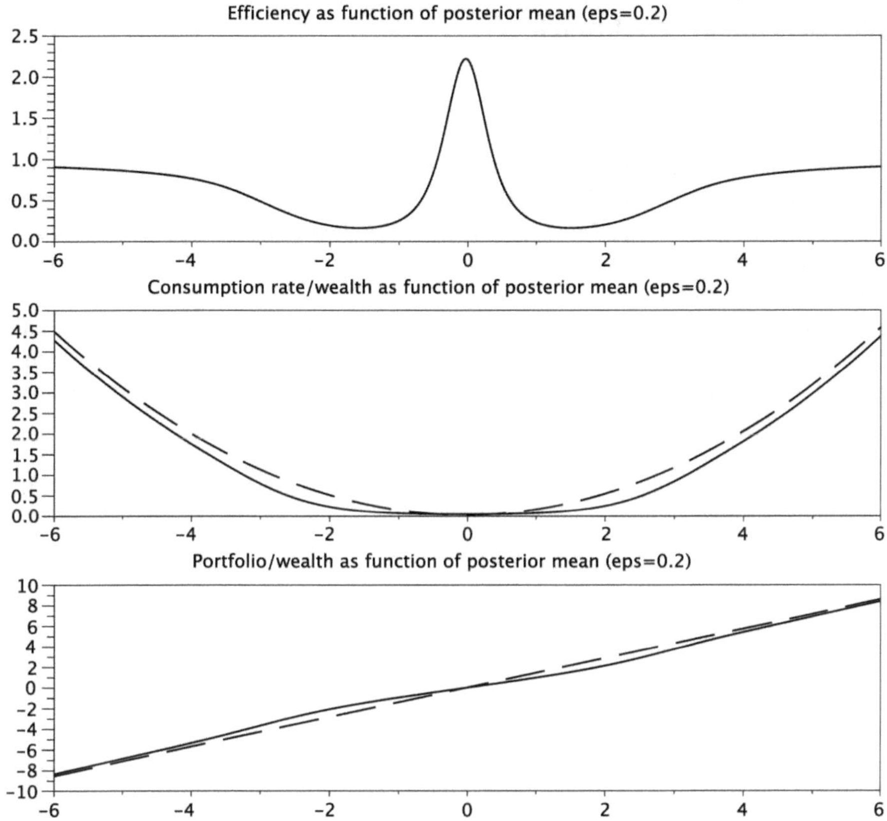

Fig. 2.22 Plots of efficiency, c/w and θ/w as a function of posterior mean for the example of Section 2.27

that at some time in the future the growth rate will move away from zero and the stock will become more attractive. So it is better to be at zero posterior mean in the model where the growth rate can change than it would be to be at a certain zero mean which never changed. As we move to more extreme posterior means, the asset is very desirable, and is not likely to change over moderate timescales, so we see a performance not unlike what we would get with constant but extreme growth rate. The consumption and portfolio plots reinforce the message that if the posterior mean is far from zero the stock is a good buy (if $\hat{\alpha}$ is positive). The dashed plots show the values which would be obtained for the Merton problem with the corresponding fixed value of the growth rate.. Notice that with variable growth rate we find the optimal behaviour is more cautious than it would be with the same fixed growth rate—if we knew the growth rate with certainty, we would consume more rapidly, and we would take a more extreme portfolio position.

2.28 Beating a Benchmark

The idea here is that the agent has a terminal wealth objective, but he is constrained always to generate at least some multiple of a benchmark process ξ. This would be the objective of a fund manager who takes money from investors and promises that they will always get at least 70 % of the S&P500 index, for example. This constraint is expressed as $w_T \geq b\xi_T$, where $0 < b < 1$ and the benchmark process is started at the same value $\xi_0 = w_0$. We shall assume the standard wealth dynamics (2.1) without consumption.

The time-0 cost of the guarantee $b\xi_T$ is bw_0 so the manager has to set aside this much money at time 0 to buy the guarantee, and may invest freely with the remaining $x_0 = (1-b)w_0$ to generate a non-negative wealth x_T at time T. His optimization problem is therefore

$$\sup_{x_T \geq 0} E[u(b\xi_T + x_T)] \quad \text{subject to} \quad E[\zeta_T x_T] = x_0. \tag{2.185}$$

In Lagrangian form, the problem is

$$\sup_{x_T \geq 0} E\left[u(b\xi_T + x_T) + \lambda(x_0 - \zeta_T x_T)\right],$$

and the first-order conditions for the problem are

$$u'(b\xi_T + x_T) - \lambda\zeta_T \leq 0, \tag{2.186}$$

with equality when $x_T > 0$. The optimal solution x_T^* is therefore of the form

$$x_T^* = (I(\lambda\zeta_T) - b\xi_T)^+ \tag{2.187}$$

for some $\lambda > 0$ chosen to match the budget constraint.

So how would it look in an example? Suppose that we take $w_0 = 1$, and let ξ be the stock S, so that

$$\xi_T = \exp(\sigma W_T + (\mu - \tfrac{1}{2}\sigma^2)T).$$

Take u to be CRRA as we often do, $u'(x) = x^{-R}$, and then

$$I(\lambda\zeta_T) = \lambda^{-1/R} \exp\left(\frac{\kappa}{R} W_T + \frac{r + \frac{1}{2}\kappa^2}{R} T\right).$$

Notice that $\kappa/\sigma R = \pi_M$, the Merton proportion. It is reasonable to suppose that $\pi_M < 1$; we do not expect investors to go out and borrow money to put everything into the stock. In that case, a little thought shows that $I(\lambda\zeta_T) > b\xi_T$ if and only if $W_T < a$ for some a determined from the parameters of the problem. After some routine calculation, the budget constraint appears as

$$1-b=\lambda^{-1/R}\exp\left\{(1-R)(r+\tfrac{1}{2}\kappa^2)T/R+\kappa^2(1-R)^2T/2R^2\right\}\Phi\left(\frac{c+\kappa(R-1)T/R}{\sqrt{T}}\right)$$
$$-b\exp\left\{(\mu-\tfrac{1}{2}\sigma^2-r-\tfrac{1}{2}\kappa^2)T+\tfrac{1}{2}(\sigma-\kappa)^2T\right\}\Phi\left(\frac{c-(\sigma-\kappa)T}{\sqrt{T}}\right), \quad (2.188)$$

where

$$c=\left\{(r+\tfrac{1}{2}\kappa^2)T-(\mu-\tfrac{1}{2}\sigma^2)RT-\log\lambda\right\}/(\sigma R-\kappa),$$

and Φ is the standard normal distribution function.

Numerics. We see a plot of the solution in Fig. 2.23. The time horizon was $T = 1$, and the promise was to pay out at least 70 % of the benchmark. Notice that the investors will receive the benchmark if the benchmark has done reasonably well, but will exceed the benchmark if it does poorly; as expected, this fund will protect investors to some extent against a fall of the benchmark.

As a comparison, we next show how the problem and its solution would change if the fund manager promised that the *gain* in the investors' wealth would be at least 70 % of the *gain* in the S&P500 index. The solution is show in Fig. 2.24. Once again, the guarantee is what you get for extreme values of wealth, it is only in the middle range that the strategy improves on the guarantee. The range of improvement is much smaller this time than when the guarantee only promised to beat 70 % of the terminal value, but that is not surprising; this time, the lower bound as a function of wealth is a straight line of slope 0.7 passing through the point (1, 1), but in the first formulation, the guarantee was a straight line with slope 0.7 passing through (0, 0), and this is always below the value of the guarantee defined in terms of the gain.

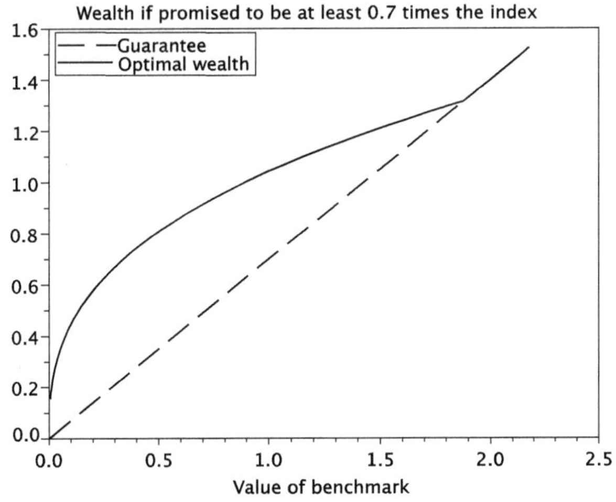

Fig. 2.23 Optimal terminal wealth as a function of the underlying benchmark value if the manager has promised to pay at least 70 % of the value of the benchmark at time T

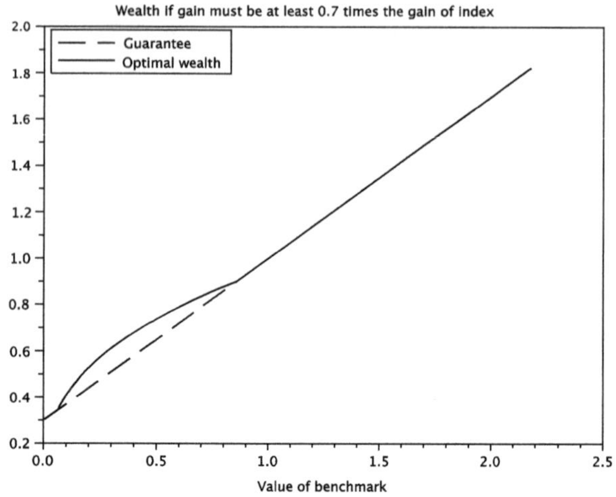

Fig. 2.24 Optimal terminal wealth as a function of the underlying benchmark value if the manager has promised that the gain in the fund will be at least 70 % of the gain in the benchmark by time T

2.29 Leverage Bound on the Portfolio

This example has been studied by Phil Dybvig and Yajun Wang. The story is a small variation of the basic Merton problem, but already this introduces features that need to be handled carefully. We suppose that the agent has the standard wealth dynamics (2.1), with the standard objective (2.2), but that the portfolio process θ is constrained:

$$\theta_t \leq a w_t \tag{2.189}$$

for some positive constant a. The HJB equation for the value function

$$V(w) \equiv \sup_{c \geq 0, \theta \leq aw} E\left[\int_0^\infty e^{-\rho t} u(c_t)\, dt \,\middle|\, w_0 = w\right] \tag{2.190}$$

is just the analogue of the familiar HJB equation but with a constraint on the portfolio variable:

$$0 = \sup_{c \geq 0,\, \theta \leq aw} \left[-\rho V + u(c) + (rw + \theta(\mu - r) - c)V' + \tfrac{1}{2}\sigma^2\theta^2 V''\right]. \tag{2.191}$$

If we were to suppose that u is CRRA to allow us to use some scaling, then we have assumed away all the interesting behaviour: the ratio θ_t / w_t would be the constant π_M for the Merton investor, and now for the constrained investor the best that can be done will be to take $\theta_t / w_t = \min\{a, \pi_M\}$.

So we are forced to consider other utilities with variable coefficient of relative risk aversion. If we take

$$u(x) = \frac{x^{1-R_1}}{1-R_1} + A\,\frac{x^{1-R_2}}{1-R_2} \tag{2.192}$$

for some positive constants $R_1 < R_2$, then we have an investor who for large values of wealth behaves like a CRRA investor with coefficient R_1 of relative risk aversion, but for small values of wealth he behaves like a more cautious investor with coefficient R_2 of relative risk aversion.

Numerics. In the numerical example, $R_1 = 1.2$ and $R_2 = 2.5$, with $A = 1$, and $K = 0.469$. The plots of the portfolio divided by wealth, and of consumption divided by wealth perform as we would expect. As wealth rises, and we become more risk tolerant, the fraction of wealth we invest in the risky asset rises from the (low) Merton proportion $(\mu - r)/\sigma^2 R_2$, but gets capped at the value K. The proportional rate of consumption falls as wealth rises, but continues to fall even after the portfolio has hit its bound; this is not surprising, because after all the consumption has not been pushed up to any bound, and should be free to adapt to the rising wealth (Fig. 2.25).

2.30 Soft Wealth Drawdown

The constraint on drawdown studied in Section 2.5 is arguably too severe, and in practice leads to trading which locks in losses, which is certainly not desirable. As an alternative, we might consider the standard wealth dynamics (2.1), with objective

$$V(w, \bar{w}) \equiv \sup E\left[\int_0^\infty e^{-\rho t}\left\{u(c_t) + C\left\{\left(\frac{w_t}{\bar{w}_t}\right)^{-a} - 1\right\}u(\bar{w}_t)\right\}dt \,\middle|\, w_0 = w, \bar{w}_0 = \bar{w}\right] \tag{2.193}$$

where $\bar{w}_t \equiv \sup_{0\leq s\leq t} w_s$ as before, and u is CRRA, $u'(x) = x^{-R}$, and $C > 0$. We shall also require that $(R-1)a > 0$. The effect of this objective is to penalise times when w is a small fraction of $\bar{w}$, that is, when we are experiencing large drawdown. However, the penalty is less absolute than the example of Section 2.5.

A familiar scaling argument tells us that $V(\lambda w, \lambda \bar{w}) = \lambda^{1-R}\, V(w, \bar{w})$ for any $\lambda > 0$, and the HJB equation for this problem is just

$$0 = \sup_{c\geq 0,\,\theta}\left[-\rho V + u(c) + C\left\{\left(\frac{w}{\bar{w}}\right)^{-a} - 1\right\}u(\bar{w}) + \{rw + \theta(\mu - r) - c\}V_w + \tfrac{1}{2}\sigma^2\theta^2 V_{ww}\right] \tag{2.194}$$

along with the boundary derivative condition

$$V_{\bar{w}}(w, w) = 0 \qquad \forall w > 0. \tag{2.195}$$

We therefore have a solution of the form $V(w, \bar{w}) = \bar{w}^{1-R}\, v(x)$, where $x \equiv w/\bar{w}$. Rewriting (2.194) in terms of this gives us

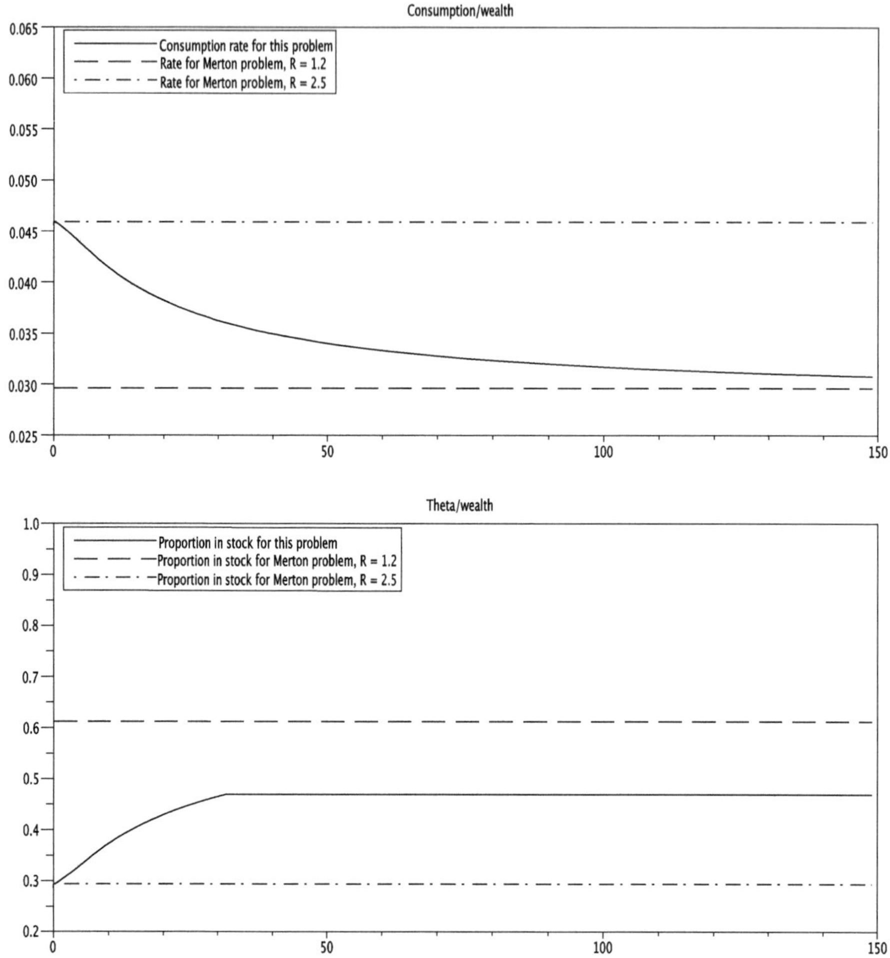

Fig. 2.25 Plots of consumption divided by wealth, and holding of the risky asset divided by wealth for the example of a leverage bound on the portfolio, Section 2.29

$$0 = \sup_{s \geq 0,\, q} \bar{w}^{1-R} \left[-\rho v + u(s) + C(x^{-a} - 1)\, u(1) + \{rx + q(\mu - r) - s\} v' + \tfrac{1}{2}\sigma^2 q^2 v'' \right] \tag{2.196}$$

along with the boundary derivative condition

$$(1 - R)\, v(1) = v'(1). \tag{2.197}$$

Optimizing in (2.196) gives us finally

$$0 = -\rho v + \tilde{u}(v') + C(x^{-a} - 1)\, u(1) + rxv' - \tfrac{1}{2}(\kappa v')^2 / v''. \tag{2.198}$$

Numerics. This problem can be solved numerically by discretizing the variable x onto a grid $x_1 < x_2 < \cdots < x_N = 1$, and using policy improvement. We have a boundary condition at $x = 1$, but it is not so clear what we should do at the lower end $x = x_1$. Everything depends on the relative sizes of $a \equiv R' - 1$ and $R - 1$. If $R' > R$, then for very low wealth levels it is only the drawdown contribution to the objective (2.193) which matters, but if $R > R'$, then the consumption contribution dominates.

In the second case, we expect that $v(x) \sim u(x)$ for very small x—that is, the value for this problem scales very much like the value for the Merton problem. If on the other hand[21] $R' \equiv a + 1 > R$, then the value for small x should scale like $v(x) \sim x^{-a}$.

Figure 2.26 shows an example of the first kind, with $a = 0.5$, $C = 10$ and default values (2.3) for all the other parameters. By contrast, Fig. 2.27 show the same plot with $a = 1.5$, $C = 10$. The dashed lines show the values which would be used in the standard Merton solution. Notice how consumption drops as w falls when we are more concerned about the effect of wealth drawdown, in Fig. 2.27. When we are more concerned about consumption effects, then the shape of the consumption curve, in Fig. 2.26, is convex. The efficiency for the first example is 0.96180, and for the second is 0.89329.

2.31 Investment with Retirement

This is a pretty example presented by Lim & Shin [25], who discuss the case of general u; as usual, we will just deal with the case of CRRA utility for simplicity of exposition.

In this example, we consider the situation of an agent who is investing in the standard market, but who is working, generating income at a fixed rate ε, with a utility penalty for working. At a moment of his choosing, the agent retires, ceases to receive his income, but also benefits by not having the disutility of working. How should he invest, and when should he choose to retire?

If τ denotes the time the agent chooses to retire, then the wealth dynamics are slightly modified from (2.1). We have instead

$$dw_t = r\, w_t\, dt + \theta_t(\sigma dW_t + (\mu - r)\, dt) + \varepsilon I_{\{t \le \tau\}}\, dt - c_t\, dt. \tag{2.199}$$

The agent's objective we shall assume is to achieve

$$V(w) \equiv \sup E\left[\int_0^\infty e^{-\rho t} \left\{ u(c_t) - \lambda I_{\{t \le \tau\}} \right\} dt \,\middle|\, w_0 = w \right]. \tag{2.200}$$

---

[21] We omit consideration of the case $R = R'$, which is a knife-edge case.

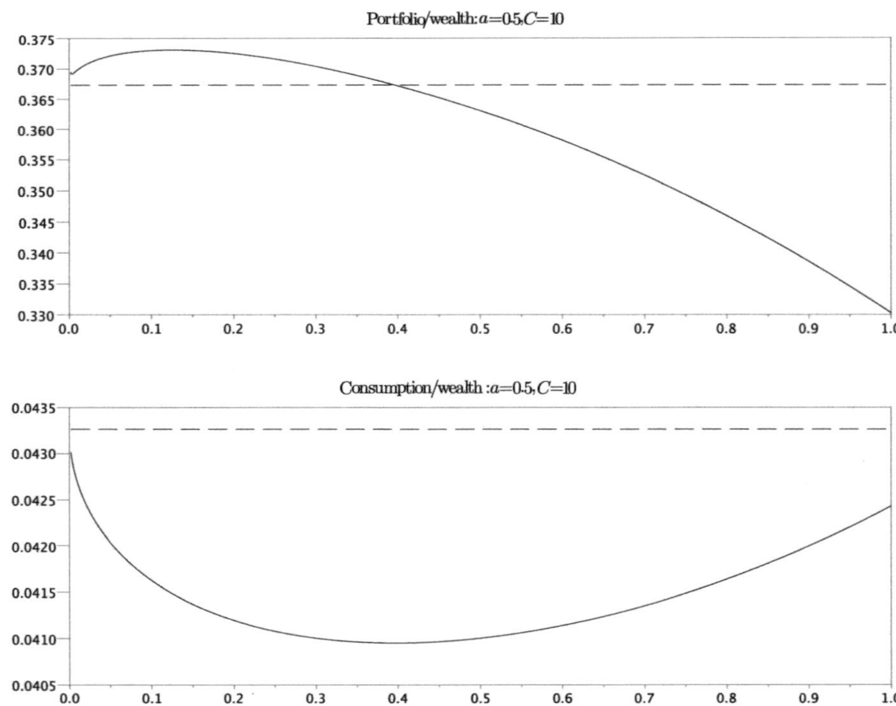

Fig. 2.26 Plots of consumption divided by wealth, and holding of the risky asset divided by wealth for the soft wealth drawdown example of Section 2.30

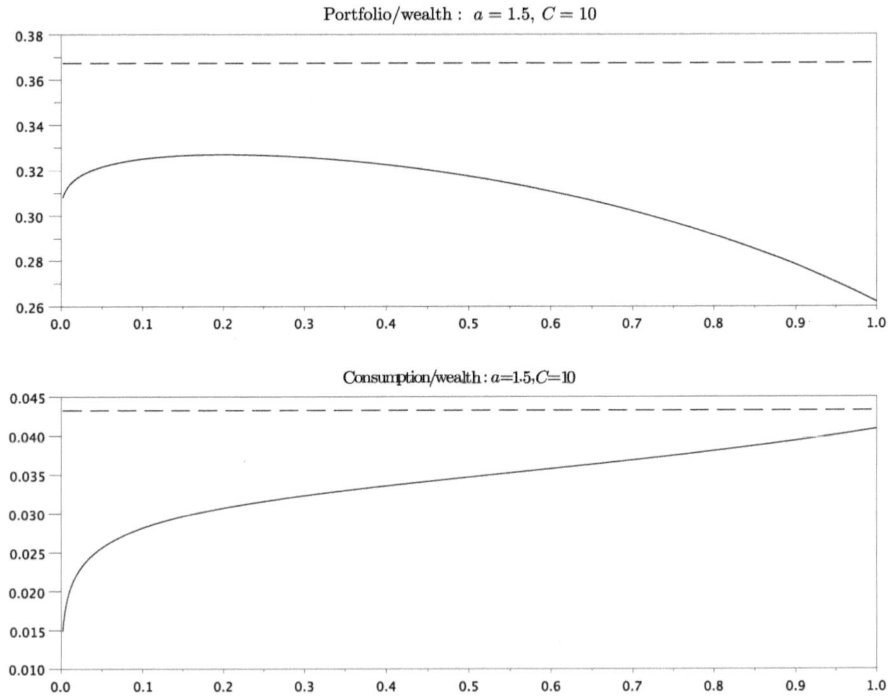

Fig. 2.27 Plots of consumption divided by wealth, and holding of the risky asset divided by wealth for the soft wealth drawdown example of Section 2.30

It is reasonable to guess that the agent's optimal policy will be to retire as soon as w reaches some critical value w^*. If this is so, then the HJB equations will be

$$\begin{aligned} 0 &= \sup\left[-\rho V + u(c) - \lambda I_{\{w \le w^*\}} + (rw + \theta(\mu - r) + \varepsilon I_{\{w \le w^*\}} - c)V' + \tfrac{1}{2}\sigma^2\theta^2 V''\right] \\ &= -\rho V + \tilde{u}(V') - \lambda I_{\{w \le w^*\}} + (rw + \varepsilon I_{\{w \le w^*\}})V' - \tfrac{1}{2}\kappa^2 \frac{(V')^2}{V''}. \end{aligned} \tag{2.201}$$

It is moreover clear that for $w \ge w^*$ we must have

$$V(w) = V_M(w) = \gamma_M^{-R} u(w), \tag{2.202}$$

because once the agent's wealth has got up to the critical value he is just a standard Merton investor. So if we just restrict attention to $w < w^*$ for now, the HJB equation (2.201) says

$$0 = -\rho V + \tilde{u}(V') - \lambda + (rw + \varepsilon)V' - \tfrac{1}{2}\kappa^2 \frac{(V')^2}{V''}.$$

This cries out for the dual variable transformation; in the customary notation, the equation for the dual value function J is

$$0 = \tilde{u}(z) - \lambda + \varepsilon z - \rho J + (\rho - r) z J' + \tfrac{1}{2}\kappa^2 z^2 J'', \tag{2.203}$$

at least in the region $z \geq z^* = V'(w^*)$. This linear second-order ODE has the explicit solution

$$J(z) = \frac{-\tilde{u}(z)}{Q(1-1/R)} - \frac{\lambda}{\rho} + \frac{\varepsilon z}{r} + Az^{-\alpha} + Bz^{\beta}, \tag{2.204}$$

where $-\alpha < 0 < 1 < \beta$ are the roots of the quadratic Q defined at (1.50). In order that J remains concave and monotone decreasing for very large z, it has to be that $B = 0$, and so we have that J is defined by[22]

$$J(z) = \gamma_M^{-1}\, \tilde{u}(z) \qquad (z \leq z^*) \tag{2.205}$$

$$= \gamma_M^{-1}\, \tilde{u}(z) - \frac{\lambda}{\rho} + \frac{\varepsilon z}{r} + A\left(\frac{z}{z^*}\right)^{-\alpha} \qquad (z \geq z^*) \tag{2.206}$$

for some z^* and A chosen to make J defined by (2.205), (2.206) to be C^1 at z^*. Solving the equations gives us explicitly that

$$z^* = \frac{\lambda r \alpha}{\varepsilon \rho (1+\alpha)}, \qquad A = \frac{\lambda}{\rho(1+\alpha)}. \tag{2.207}$$

The critical value of wealth is now given by

$$w^* = -J'(z^*) = \gamma_M^{-1}\, (z^*)^{-1/R}. \tag{2.208}$$

2.32 Parameter Uncertainty

The dynamics of wealth are as usual

$$dw_t = rw_t dt + \theta_t \cdot \sigma\{dW_t + (\alpha - r\sigma^{-1}\mathbf{1})dt\} - c_t dt \tag{2.209}$$

which we have written in a slightly unusual way, because we intend now to suppose that the parameter α is *not* known with certainty, rather that we shall have a prior $N(\hat{\alpha}_0, \tau_0^{-1})$ distribution for it. The volatility matrix σ is $n \times n$, and assumed known and non-singular.

This means that we shall have to filter the value of α from the observed price of the stock. Thus we see the processes $\log S_t^i / S_0^i = \sum_j \sigma_{ij}(W_t^j + \alpha^j t) - \frac{1}{2} v_{ii} t$, or equivalently the processes $X_t^j \equiv W_t^j + \alpha^j t$, and must filter α from that.

---

22 We used the easily-verified fact that $Q(1 - 1/R) = -\gamma_M$.

The slick way to do this is to write down the likelihood for a path $(X_s)_{0\le s\le t}$ with respect to Wiener measure:

$$\exp(\alpha\cdot X_t-\tfrac{1}{2}|\alpha|^2 t), \tag{2.210}$$

according to the Cameron-Martin-Girsanov theorem. Multiplying by the prior density of α gives us the posterior for α given $(X_s)_{0\le s\le t}$, which is proportional to

$$\exp\Big[\alpha\cdot X_t-\tfrac{1}{2}|\alpha|^2 t-\tfrac{1}{2}(\alpha-\hat{\alpha}_0)\cdot\tau_0(\alpha-\hat{\alpha}_0)\Big]\propto\exp\Big[-\tfrac{1}{2}(\alpha-\hat{\alpha}_t)\cdot\tau_t(\alpha-\hat{\alpha}_t)\Big],$$

where

$$\tau_t\equiv\tau_0+tI, \tag{2.211}$$
$$\hat{\alpha}_t\equiv\tau_t^{-1}(\tau_0\hat{\alpha}_0+X_t). \tag{2.212}$$

We see that the posterior for α is again multivariate Gaussian. It is a simple result of filtering theory (see, for example, [34], VI.8) that the observation process X can be expressed as

$$\begin{aligned}dX_t&=dW_t+\alpha dt\\&=d\hat{W}_t+\hat{\alpha}_t dt,\end{aligned} \tag{2.213}$$

where $\hat{W}$ is a martingale in the observation filtration $\mathscr{G}_t\equiv\sigma(\{X_u:0\le u\le t\})$. Observing that the quadratic variation process of X is t, we see that $\hat{W}$ is actually a Brownian motion. Now X and $\hat{\alpha}$ are related via (2.212), so applying integration-by-parts, we deduce the key relation

$$d\hat{\alpha}_t=\tau_t^{-1}d\hat{W}_t. \tag{2.214}$$

It should not be a surprise that the finite-variation parts vanish, since $\hat{\alpha}_t=E\left[\alpha\mid\mathscr{G}_t\right]$ is a martingale.

If we now switch to the filtration $(\mathscr{G}_t)$, the wealth dynamics (2.209) gets changed to

$$dw_t=rw_t dt+\theta_t\cdot\sigma\{d\hat{W}_t+(\hat{\alpha}_t-r\sigma^{-1}\mathbf{1})dt\}-c_t dt.$$

But we know how to proceed to solve this sort of problem; we find the state-price density process, and express the solution in terms of it. In this instance, the state-price density process satisfies

$$\zeta_t^{-1}d\zeta_t=-rdt+(r\sigma^{-1}\mathbf{1}-\hat{\alpha}_t)d\hat{W}_t, \tag{2.215}$$

since this is what discounts at the riskless rate, and changes the rate of growth of the risky assets to r. We now abbreviate $\kappa_t\equiv\hat{\alpha}_t-r\sigma^{-1}\mathbf{1}$ and notice that

$$d\kappa_t = \tau_t^{-1} d\hat{W}_t. \tag{2.216}$$

Looking at (2.215), we see that we need to simplify

$$\begin{aligned}
\kappa_t d\hat{W}_t &= \kappa_t \cdot \tau_t d\kappa_t \\
&= d\{\tfrac{1}{2}\kappa_t \cdot \tau_t\, \kappa_t\} - \tfrac{1}{2}|\kappa_t|^2 dt - \tfrac{1}{2}\mathrm{tr}(\tau_t^{-1})dt.
\end{aligned}$$

We may now re-express the state-price density much more simply:

$$\begin{aligned}
\zeta_t &= \exp\Big[- rt - \tfrac{1}{2}\kappa_t \cdot \tau_t\, \kappa_t + \tfrac{1}{2}\kappa_0 \cdot \tau_0\, \kappa_0 + \int_0^t \tfrac{1}{2}\mathrm{tr}(\tau_s^{-1})ds \Big] \\
&= \left\{ \frac{\det \tau_t}{\det \tau_0} \right\}^{1/2} \exp\big[-rt - \tfrac{1}{2}\kappa_t \cdot \tau_t\, \kappa_t + \tfrac{1}{2}\kappa_0 \cdot \tau_0\, \kappa_0\big].
\end{aligned} \tag{2.217}$$

Expressing optimal consumption in terms of ζ, we have

$$e^{-\rho t} u'(c_t^*) = \lambda_0 \zeta_t$$

for some $\lambda_0 > 0$ which is determined by the budget equation

$$\begin{aligned}
w_0 &= E\left[\int_0^\infty \zeta_s c_s^*\, ds \right] \\
&= \lambda_0^{-1/R} E\left[\int_0^\infty e^{-\rho s/R} \zeta_s^{1-1/R}\, ds \right] && (2.218) \\
&\equiv \lambda_0^{-1/R} \varphi(\hat{\alpha}_0, \tau_0), && (2.219)
\end{aligned}$$

say. The optimised objective is

$$\begin{aligned}
E \int_0^\infty e^{-\rho t} u(c_t^*) dt &= \frac{\lambda_0^{1-1/R}}{1-R} E\left[\int_0^\infty e^{-\rho s/R} \zeta_s^{1-1/R}\, ds \right] \\
&= \frac{\lambda_0^{1-1/R}}{1-R}\, \varphi(\hat{\alpha}_0, \tau_0) \\
&= u(w_0)\, \varphi(\hat{\alpha}_0, \tau_0)^R.
\end{aligned} \tag{2.220}$$

The extent to which we may express the solution to this problem explicitly depends on the extent to which we can simplify the expression for φ. We can go quite far, but not all the way. The integral expression (2.219) shows that we will need a simpler expression for $E\zeta_t^b$, where $b = 1 - R^{-1}$ in this case. The variable ζ_t is the exponential of a squared Gaussian, so we are able to compute the required expectation in closed form. After some calculations, we obtain finally

$$E\zeta_t^b = \left(\frac{\det \tau_0}{\det(bt+\tau_0)}\right)^{1/2}\left(\frac{\det \tau_t}{\det \tau_0}\right)^{b/2} \exp\left\{-\frac{tb(1-b)}{2}\ \kappa_0\cdot\tau_0(bt+\tau_0)^{-1}\kappa_0 - rbt\right\}. \tag{2.221}$$

To evaluate φ, we have to integrate (2.221) with respect to t; while this is easy enough to do numerically, it cannot be done in closed form. Nevertheless, if all that we are concerned with is the Merton *wealth* problem (that is, maximising the expected utility of wealth at time T), then (2.221) is all we need, and the problem can be done entirely explicitly.

Writing $\lambda_t \equiv \lambda_0\zeta_t$ and thinking what the budget equation (2.219) becomes at time t, we see that

$$c_t^* = \frac{w_t}{\varphi(\hat{\alpha}_t, \tau_t)}, \tag{2.222}$$

$$w_t = e^{-\rho t/R}\lambda_t^{-1/R}\varphi(\hat{\alpha}_t, \tau_t), \tag{2.223}$$

$$\theta_t^* = R^{-1}\sigma^{-2}(\sigma\hat{\alpha}_t - r\mathbf{1}) + \sigma^{-1}\tau_t^{-1}\nabla\log\varphi(\hat{\alpha}_t, \tau_t), \tag{2.224}$$

this last coming from expanding w_t by Itô's formula, and matching the coefficient of $d\hat{W}$.

Observe that the optimal portfolio consists of two terms, the first being the Merton proportion when the posterior mean for α is substituted for the (true, supposed-known, value), the second of which is the alteration required to account for the fact that the mean is not known precisely. Notice that as $t \to \infty$, this second term goes to zero (some checking of the properties of φ is needed to decide this).

What about the efficiency of the Merton investor who faces uncertainty in the value of α? Let us take some typical values for the parameters in the case of a single risky asset, and see what we get.

Taking $r = 0.05$, $\sigma = 0.25$, $\hat{\alpha}_0 = 0.56$, $\rho = 0.02$, $R = 2$ and $\tau_0 = 10$, we find that efficiency *drops to 73.19 % !* The initial proportion that should be invested in the risky asset changes from 73.37 % in the standard Merton problem to 40.96 % once we take account of parameter uncertainty, another substantial difference. The rate at which we consume initially is 4.36 % of wealth, in contrast to the 5.10 % of wealth that the standard Merton investor would follow!

Let us look at one final question before finishing with our study of the effects of uncertainty about α, and that is to understand what would happen if we faced parameter uncertainty, but just used the naive policy of investing and consuming according to the standard Merton rule, simply substituting in our posterior mean for α at time t as if it were known and fixed. For simplicity, let us restrict to the case of a single risky asset.

The effect of this is that we hold proportion

$$\hat{\pi}_t = \frac{\sigma\hat{\alpha}_t - r}{\sigma^2 R}$$

of our wealth in the risky asset at time t, and are consuming at rate

$$\hat{\gamma}_t = R^{-1}\big[\rho + (R-1)(r + \tfrac{1}{2}\sigma^2 R\hat{\pi}_t^2)\big]$$

at time t. The wealth dynamics are

$$dw = rwdt + \hat{\pi} w\sigma(d\hat{W} + (\hat{\alpha} - r/\sigma)dt) - \hat{\gamma} wdt$$

so that

$$w^{-1}dw = \sigma\hat{\pi} d\hat{W} + \{(r-\rho)/R + \tfrac{1}{2}\sigma^2\hat{\pi}^2(R+1)\}dt,$$

after some calculations. As before, the stochastic integral term can be simplified:

$$\sigma\hat{\pi} d\hat{W} = d(\tfrac{1}{2}\sigma^2 R\hat{\pi}^2\tau_t) - (2R\tau_t)^{-1}dt - \tfrac{1}{2}\sigma^2 R\hat{\pi}^2 dt,$$

which leads to the expression

$$w_t = w_0 \exp\big[\tfrac{1}{2}\sigma^2 R\hat{\pi}_t^2\tau_t - \tfrac{1}{2}\sigma^2 R\hat{\pi}_0^2\tau_0\big]e^{-(\rho-r)t/R}\left(\frac{\tau_0}{\tau_t}\right)^{1/2R} \tag{2.225}$$

for the wealth process. The value of the objective is

$$E\int_0^\infty e^{-\rho t}u(\hat{\gamma}_t w_t)\,dt,$$

and this can be evaluated numerically at least. When we do this for the numerical example studied above, we find that this naive policy achieves an efficiency of 72.61 %, *hardly any lower than the optimum achieved by the investor who adjusts his portfolio and consumption proportions according to the full Bayesian analysis!*

The message from this example is that pretending that we know α may not lead us to follow rules which are suboptimal by very much; however, it will lead us to be grossly over-optimistic about how well we are doing.

2.33 Robust Optimization

The title of this section is arguably an oxymoron; if we have optimized, then it would have to be with respect to a specific model, whereas the essence of robustness is that our conclusions should be insensitive to precise modelling assumptions.

Let us take an example where we have the standard wealth dynamics (2.1) and the standard objective (2.2), but the growth rate μ is not supposed known; all we shall assume is that $a \leq \mu \leq b$ for some[23] $a \leq r \leq b$. If the (Merton) investor knows the value of μ, then he follows the optimal policy of investing the Merton proportion $\pi_M = (\mu - r)/\sigma^2 R$ of his wealth in the risky asset, and consuming at rate $\gamma_M w_t$, where

$$\gamma_M = \{\rho + (R-1)(r + \kappa^2/2R)\}/R.$$

The value he achieves is then given by (see (1.30))

$$V_M(w) = \gamma_M^{-R} U(w).$$

Now the term 'robust' is often interpreted to mean 'minimax', which is to say that an opponent chooses which probability model from a pre-specified set will be used, with the aim of making your value as small as possible. So in this setting we have the problem of

$$\begin{aligned}\inf_{a \leq \mu \leq b} \sup_{(n,c) \in \mathscr{A}(w)} \Psi(n, c; \mu) &\equiv \inf_{a \leq \mu \leq b} \sup_{(n,c) \in \mathscr{A}(w)} E^\mu \left[\int_0^\infty e^{-\rho t} u(c_t)\, dt \,\middle|\, w_0 = 0 \right] \\ &= \inf_{a \leq \mu \leq b} \gamma_M^{-R} U(w).\end{aligned}$$

Inspection of the explicit form of γ_M reveals that the best choice for your opponent is to pick $\mu = r$, resulting in $\kappa = 0$. If this is the value of μ, then $\pi_M = 0$ and you invest all of your wealth only in the bank account. The minmax inequality

$$\inf_{a \leq \mu \leq b} \sup_{(n,c) \in \mathscr{A}(w)} \Psi(n, c; \mu) \geq \sup_{(n,c) \in \mathscr{A}(w)} \inf_{a \leq \mu \leq b} \Psi(n, c; \mu) \tag{2.226}$$

clearly holds with equality when on the right-hand side the policy chosen is to invest nothing in the risky asset, and to consume at the rate $\gamma_M^0 w_t$, where

$$\gamma_M^0 \equiv \{\rho + (R-1)r\}/R;$$

compare with the definition of γ_M. If you choose to use that policy, then it does not matter what drift μ your opponent chooses!

Thus in this situation, the minimax solution is for you to put *nothing* in the risky asset, and this is very typical of minimax solutions; they are generally over-cautious.

So what could we do instead? If we are to consider the performance of an investment strategy faced with a set of possible alternative models, a Bayesian approach has always seemed to me to be more attractive than a minimax approach, and our earlier example of Section 2.32 presents such an analysis. Other than this, we may try to resort to some *intelligent heuristic*. Here is an example.

---

[23] The assumption that $a \leq r \leq b$ is merely for expositional convenience. You are invited to work out what happens if this condition does not hold.

Suppose that you have N advisors, each of whom thinks that the (d-dimensional) log-price vector $X_t \equiv \log S_t$ of some asset is a Lévy process in $\mathbb{R}^d$. These advisors may invest in a riskless bank account, or in the assets; at time 0, you split your initial wealth 1 among the advisors, entrusting advisor j with initial wealth w_0^j. Suppose that advisor j has objective

$$V_j(w) = \sup E^j[u(w_T)] \tag{2.227}$$

for some (large) T, where E^j is expectation with respect to advisor j's probability P^j, which we assume is given by a density Λ_T^j with respect to some reference probability P. We shall also assume that u is CRRA, so that the optimal investment for advisor j would be to put fixed fractions π^j of wealth into the risky assets—a so-called *fixed-mix* rule. Now assuming that the different advisors have a common[24] state-price density process ζ, it would have to be that the optimal wealth process w^j for advisor j would satisfy the relation

$$\Lambda_T^j \, u'(w_T^j) = \alpha_j \, \zeta_T \tag{2.228}$$

for some constant α_j. Turning this around, and using the fact that u is CRRA, we learn that

$$\Lambda_T^j = \alpha_j \, \zeta_T \, (w_T^j)^R. \tag{2.229}$$

Now this is an intriguing relation, because it tells us that (apart from the constants α_j) the relative degrees of belief in the different advisors' modelling hypotheses at time *T are proportional to* $(w_T^j)^R$, that is, *proportional to the* Rth *powers of the wealth the advisors generated by their fixed-mix investment strategies.* To simplify matters, let us now suppose that $R = 1$; *all of your advisors (and you) have log preferences.* Taking expectations on both sides of (2.229) reveals that

$$1 = \alpha_j E\left[\zeta_T w_T^j\right] = \alpha_j w_0^j,$$

so that $\alpha_j = 1/w_0^j$.

If you started at time 0 with prior beliefs (p_j) in the different advisors (that is, you initially believed that advisor j had the correct model with probability p_j), then at time T your beliefs about the true model are summarized in your likelihood-ratio martingale

$$\bar{\Lambda}_T = \sum_j p_j \Lambda_T^j = \zeta_T \sum_j p_j \alpha_j \, w_T^j. \tag{2.230}$$

Assuming that you also share the same state-price density process ζ, your optimal wealth $\bar{w}_T$ at time T would satisfy the analogue of (2.228), namely,

---

24 This assumption would be correct if the Lévy process was a Brownian motion with drift, when the market is complete, but is otherwise a big ask.

$$\bar{\Lambda}_T u'(\bar{w}_T) = \bar{\Lambda}_T / \bar{w}_T = \beta \zeta_T,$$

from which using (2.230) we discover that

$$\bar{w}_T = \sum_j p_j \alpha_j w_T^j = \sum_j p_j \frac{w_T^j}{w_0^j}. \tag{2.231}$$

This simple statement reveals two interesting consequences: firstly, we just sit back and let the advisors work without any interference; and secondly, the distribution of our wealth among the available assets is according to the averaged fix-mix rule $\bar{\pi}$ satisfying

$$\bar{w}_t \bar{\pi}_t = \sum_j \frac{p^j w_t^j}{w_0^j} \pi^j; \tag{2.232}$$

that is, *we weight the portfolio choice* π^j *of advisor* j *according to his current contribution to our overall wealth!*

Now we can see the shape of a method emerging. under the original hypothesis that each advisor believes that the assets are log-Lévy, we would have to look at each advisor's assumed model, and compute the corresponding π^j; but in fact, all that matters at the aggregate level is *what* π^j *the advisors used*, not what log-Lévy model they assumed. So we simply need to consider a set of fixed-mix rules, and weight our investment according to how well those fixed-mix rules performed up to the current time. The next step would be to consider the set of *all possible* fixed-mix rules, and weight according to how well they had done up to the current time; and the example of Section 2.32 does exactly that in the situation with log Brownian assets and a Gaussian prior over the growth rates. In more detail, for a log investor with a finite time horizon, the wealth process w_t is proportional to ζ_t^{-1}, where ζ_t is given by (2.217). From the dynamics (2.216) of κ_t, we deduce after some calculations that the log investor will invest proportionally to wealth at time t with the weights

$$\pi_t = (\sigma^T)^{-1} \kappa_t = (\sigma \sigma^T)^{-1} (\hat{\mu}_t - r\mathbf{1}).$$

To simplify the discussion we now suppose that $r = 0$, $\tau_0 = 0$. Remembering that $\mu = \sigma \alpha$, the optimal portfolio weights at time t become

$$\pi_t = (\sigma \sigma^T)^{-1} \hat{\mu}_t = (\sigma^T)^{-1} \hat{\alpha}_t. \tag{2.233}$$

Now an advisor who believes that the true value of α is a will invest according to the fixed-mix rule with proportions $p = (\sigma \sigma^T)^{-1} \mu = (\sigma^T)^{-1} a$. This advisor will generate wealth

$$w_t^a = \exp\left\{p \cdot \sigma(W_t + \alpha t) - \tfrac{1}{2} |\sigma^T p|^2 t\right\} = \exp\left\{a \cdot X_t - \tfrac{1}{2} |a|^2 t\right\} \tag{2.234}$$

by time t. If we follow the course of action determined by the rough argument just outlined, we should weight the advisors according to the outcomes w_t^a of their fixed-mix investments, which would mean that we weight the beliefs about a according to the posterior Gaussian distribution with mean $X_t/t = \hat{\alpha}_t$. Weighting the portfolio choices of the advisors according to this distribution would mean that we use portfolio proportions equal to the mean of $p = (\sigma^T)^{-1} a$ under this posterior, namely, $\pi_t = (\sigma^T)^{-1} \hat{\alpha}_t$. In other words, in this special (but interesting) situation, *the rough argument leads us to carry out the optimal investment.*

There is another natural thing we could do in this situation, and that would be to consider the wealths w_t^a that would have arisen from all possible fixed-mix rules, pick the best one at time t, and then follow the recommendation of that advisor. This is the approach of Cover's universal portfolio algorithm [9]. Cover presents this approach as an ansatz, without any supporting modelling background; but if we look at the form of (2.234), we see that following the advice of the current best advisor would lead us to choose $a = \hat{\alpha}_t$. For a Gaussian distribution, the mean and the mode are the same, so the universal portfolio algorithm agrees here with the true optimum.

2.34 Labour Income

In this section, we suppose that the agent can not just invest and consume, but may also work for a fixed wage rate $a > 0$. His wealth dynamics now become

$$dw_t = rw_t\,dt + \theta(\sigma dW_t + (\mu - r)\,dt) + aL_t\,dt - c_t\,dt, \tag{2.235}$$

where $L_t \geq 0$ is the rate of working. We suppose that the agent's objective will be to obtain

$$V(w) = \sup E\left[\int_0^\infty e^{-\rho t}\,u(c_t, L_t)\,dt \,\middle|\, w_0 = w\right]. \tag{2.236}$$

The utility function u is supposed to be concave, increasing in c and decreasing in L. As usual, we can apply the Martingale Principle of Optimal Control, and derive the HJB equation for this problem:

$$0 = \sup_{c,L,\theta}\left[-\rho V + u(c, L) + \{rw + \theta(\mu - r) + aL - c\}\,V_w + \tfrac{1}{2}\sigma^2\theta^2 V_{ww}\right]. \tag{2.237}$$

Previously we would have made the problem easier by assuming some scaling properties, but this is not really possible in this situation. Nevertheless, the problem is not so very far away from those we have considered to date; if we define

$$\tilde{u}(\lambda) \equiv \sup_{c,L}\{u(c, L) + \lambda(aL - c)\} \tag{2.238}$$

then clearly $\tilde{u}$ is a convex function (though not in general increasing), and we can rewrite the HJB equation as

$$0 = -\rho V + \tilde{u}(V_w) + rwV_w - \tfrac{1}{2}\kappa^2 \frac{V_w^2}{V_{ww}}. \tag{2.239}$$

This is in a form to which we can apply the dual variable transformation $z = V_w$, $J(z) = V(w) - zw$, to give the second-order linear ODE

$$0 = \tilde{u}(z) - \rho J + (\rho - r)zJ' + \tfrac{1}{2}\kappa^2 z^2 J''. \tag{2.240}$$

The extent to which we can solve this depends now on the form of $\tilde{u}$ and any special properties this function may have. In general, we can use the representation discussed in Section 2.8 for the dual value function. However, we can also use the static optimization approach of Section 1.4, as we shall now show.

As a simple example, we propose the form

$$u(c, L) = \frac{c^{1-R}}{1-R} - AL^b \tag{2.241}$$

for some constants $R, b > 1$ and $A > 0$. The agent is going to choose to consume the stream $c_t - aL_t$, which must satisfy the budget constraint

$$E\left[\int_0^\infty \zeta_t(c_t - aL_t)\, dt\right] = w_0. \tag{2.242}$$

The aim is to maximize the objective (2.236) subject to this constraint, so by setting the optimization up in Lagrangian form we discover that the conditions for optimality will be

$$u_C(c_t, L_t) = \lambda e^{\rho t}\zeta_t, \qquad u_L(c_t, L_t) = -a\lambda e^{\rho t}\zeta_t \tag{2.243}$$

for some Lagrange multiplier $\lambda > 0$ chosen to match the budget constraint (2.242). This conclusion is generic; but for the simple special case under study here, the marginal utilities u_C and u_L are simply powers of c and L respectively, so we are able to express

$$c_t = \left(\lambda e^{\rho t}\zeta_t\right)^{-1/R}, \qquad L_t = \left(\frac{a\lambda e^{\rho t}\zeta_t}{Ab}\right)^{1/(b-1)}. \tag{2.244}$$

Introducing the abbreviation

$$h(\nu, q) \equiv E\int_0^\infty e^{-\nu t}\zeta_t^q\, dt = \left(\nu + rq + \tfrac{1}{2}\kappa^2 q(1-q)\right)^{-1}, \tag{2.245}$$

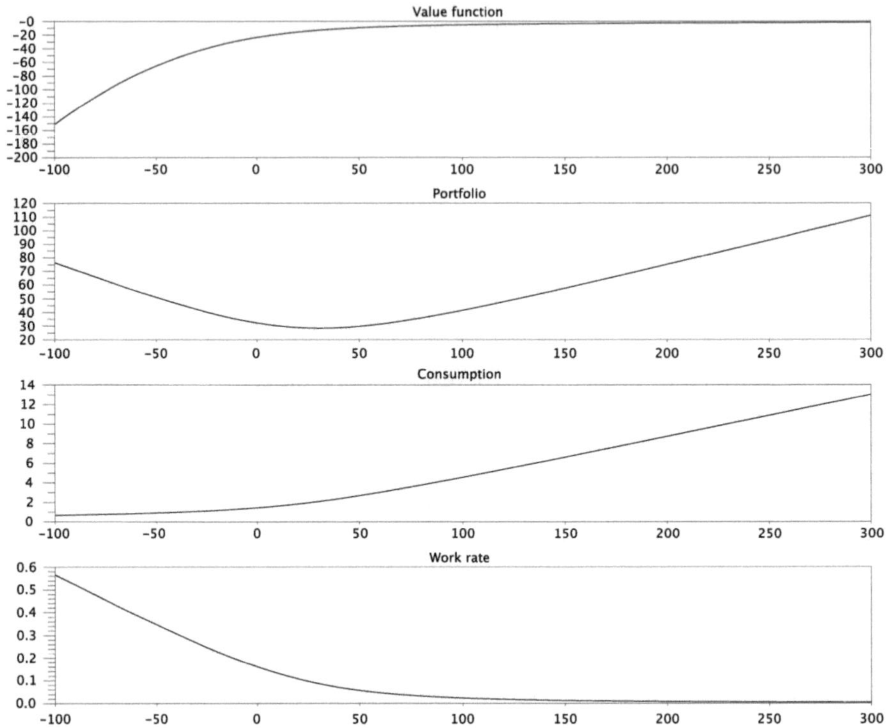

Fig. 2.28 Plots of the value, portfolio in the risky asset, and consumption rate, and rate of working for the labour income example of Section 2.34. The constants used were $A = 10$, $a = 5$, and $b = 2.2$

the budget constraint becomes

$$w_0 = \lambda^{-1/R}\, h(\rho/R, 1-R^{-1}) - \left(\frac{a\lambda}{Ab}\right)^{1/(b-1)} a\, h(-\rho/(b-1), b/(b-1)), \quad (2.246)$$

and the objective is

$$V(w_0) = \frac{\lambda^{1-1/R}}{1-R} h(\rho/R, 1-R^{-1}) - A\left(\frac{a\lambda}{Ab}\right)^{b/(b-1)} h(-\rho/(b-1), b/(b-1)). \quad (2.247)$$

Of course, in order that the integral defining $h(\nu, q)$ is well defined we shall have to have that $\nu + rq + \frac{1}{2}\kappa^2 q(1-q) > 0$ which raises a question about $h(-\rho/(b-1), b/(b-1))$; this is only going to be well defined if $-\rho/(b-1) + rq + \frac{1}{2}\kappa^2 q(1-q) > 0$, where we write q for $b/(b-1)$. A little rearrangement turns this into the condition $Q(q) < 0$, which is equivalent to saying that $q \equiv b/(b-1) < \beta$, where β is the larger root of the quadratic Q defined at (1.50).

This is not a surprising condition to demand; it tell us that unless b is large enough, the problem is ill posed. What happens if b is too small is that the penalty for working does not get large sufficiently rapidly to stop the agent working arbitrarily hard, to gain arbitrarily large consumption. Surely no-one could argue with that.

Figure 2.28 shows the form of the optimal solution. The range for wealth includes negative values, since the agent has the possibility to work very hard to recover from debt; this may be a slightly unrealistic assumption, but that is what the model gives us. It has features in common with the situation of Section 2.23 where utility was bounded below. The plots show that as the agent gets more wealthy, he consumes more, and works less, and indeed once his level of wealth gets high enough he effectively stops working. At these high wealth levels then, the agent will behave rather like a Merton investor, and we see the portfolio and consumption rates growing linearly there as we would expect. What is perhaps a little surprising is that for negative values of the wealth the agent will choose to increase his investment in the risky asset. This may make some kind of sense; he is having to work very hard, and consume little, so he is willing effectively to borrow a lot to avail himself of the superior rate of return on the risky asset.

Chapter 3
Numerical Solution

Abstract The third chapter of the book presents the main numerical methods that are useful in calculating solutions to the optimal control problems of earlier chapters when analytic methods fail. The main technique is policy improvement, but this requires an effective translation of an optimization problem for a controlled diffusion into an optimization problem for a controlled Markov chain, and various techniques for this are discussed.

We have seen in Chapter 1 a range of techniques for solving optimal investment/consumption problems, and in Chapter 2 a range of different variants on the basic problems, each approachable by one or other of the standard techniques. But explicit solutions are only available rarely, and at that point there is little we can do except prove general results; and numerics. This chapter is about numerical solution, but it is important to realize that this is not a theoretical account of numerical solution of PDEs; there is no discussion of rates of convergence, error estimates, solution concepts, regularity of solutions, uniqueness of solutions. This is mainly because I am not familiar with the literature of PDE, but also in part because I have not found the literature very helpful when it comes to *actually computing* a numerical solution! More generally, PDE is a theory that includes many areas of application besides probability, but I have always found that for applications that arise in probability, the methods of probability are at least as effective.

In the context of stochastic optimal control, the principal viewpoint expounded in this chapter is that we first *approximate the controlled diffusion by a controlled finite-state Markov chain*; and then we *exactly solve the control problem for that finite-state Markov chain*. The book of Kushner & Dupuis [24] takes the same overall view, but treats it much more completely. If we want to estimate how close the numerical solution is to the true solution, it is usually quite effective to study the nature of the stochastic approximation and work from that. In practice however, it is quicker (and tempting!) simply to refine the grid of the discretization, or to move the boundaries out a bit and see to what extent the solution changes; this is not conclusive, but gives

L. C. G. Rogers, *Optimal Investment*, SpringerBriefs in Quantitative Finance, 115
DOI: 10.1007/978-3-642-35202-7_3, © Springer-Verlag Berlin Heidelberg 2013

one a pretty good idea how good the numerics are. There is a limit to the amount of time one wants to spend on one example.

The main probabilistic tool is the policy improvement algorithm, explained for finite-state Markov chains in Section 3.1. This is a simple, intuitive, and robust method for iteratively calculating the value of a stochastic optimal control problem, and most of the numerical methods presented subsequently proceed by firstly approximating the original problem by an optimal control problem for a finite-state Markov chain, and then applying policy improvement in a suitable form.

Generically, the HJB equation can be written as

$$\mathscr{L}V + \sup_a \Psi(DV, a) = 0, \tag{3.1}$$

where $\Phi(\cdot, a)$ is some linear function of the vector[1] $DV \equiv (V, V', V'', \dot{V})$ of derivatives[2] of V, and a is the control variable. The operator $\mathscr{L}$ is a linear differential operator which does not depend on the control variable. In most of the examples studied in Chapter 2, Ψ did not depend on $\dot{V}$; the form of the problem was time-homogeneous. For such problems, we approach the approximation of the controlled diffusion (in one dimension firstly, Section 3.2) by approximating the state space by a finite grid of points, and then setting up the jump intensities of a continuous-time Markov chain to mimic the evolution of the controlled diffusion. This one-dimensional approach can be extended to higher dimensions, as is explained in Section 3.3.

This approach is not suitable for problems where the value function depends on time also. The Markov chain approximations constructed in Sections 3.2 and 3.3 accurately approximate the probabilities of jumping to nearby states, and also the *mean* times to jump to nearby states, but if the value function depends on time also, we really have to know *when* the chain jumped. Different methods are therefore required. If we consider the situation where there is no control, so (3.1) takes the form

$$\mathscr{L}V + \Psi(DV) = 0,$$

we are looking at a *parabolic* PDE. If we consider the simplest possible parabolic PDE, the one-dimensional heat equation

$$\tfrac{1}{2}V'' + \dot{V} = 0 \tag{3.2}$$

with some terminal condition $V(T, \cdot) = g(\cdot)$ given, then the most obvious Markov chain approximation to the (space-time) Brownian motion is a symmetric simple random walk on some grid with time step Δt and space step Δx. This certainly works, but is not a very good numerical method; to match the moments of the Brownian

[1] We explain the method assuming that the state variable is one-dimensional, but the methodology works also in higher dimensions.

[2] For an infinite horizon problem, the value is not time dependent, so the time derivative $\dot{V}$ of V does not appear.

motion, we have to have $\Delta t = O(\Delta x^2)$, so if we want a (modest) resolution of the order of 10^{-3} in the spatial variable, we will need millions of time-steps to compute the answer. The method of choice for (3.2) is the famous Crank-Nicolson implicit finite-difference scheme, which is globally stable, and all in all a fantastically good numerical method. If we have to use Crank-Nicolson to solve the parabolic PDE without any control element, we should expect to look for something similar to the Crank-Nicolson scheme when we have a controlled diffusion where time enters explicitly into the value function; and indeed this is what we find. For a controlled diffusion, the natural analogue of the Crank-Nicolson scheme is quite a bit more complicated, as will be explained in Section 3.4. While the solution of the heat equation by Crank-Nicolson reduces to a sequence of solutions of sparse systems of linear equations, something more intricate is going on with control. Each step of the (dynamic programming) calculation has to be solved by a recursive method. The paper of Forsyth & Labahn [16] presents and compares the classical successive over-relaxation method, and the use of policy improvement, and concludes that the second is better for stochastic control problems. We will give a very abbreviated account of [16] in Section 3.4 and refer the reader to the original paper for a thorough treatment.

The next section of this chapter, Section 3.5, discusses treatment of boundary conditions. This is usually a delicate issue in stochastic optimal control problems, and in most calculations you will spend longer getting the boundary conditions right than you will in getting the PDE right in the interior of the region. In a sentence, the recommendation of Section 3.5 is to tell a good probabilistic story at the boundaries. This is backed up with some examples of the kinds of story that could be told, and the kinds of boundary conditions which result.

The chapter finishes in Section 3.6 with a rather superficial discussion of some numerical approaches which are based on bluntly discretizing the differential equation. What is presented there is quick and dirty, quite often does not work for reasons which I hope others can illuminate, but is worth knowing about because it is usually a lot simpler to code than a full policy improvement story, and when it works it is fast and accurate.

3.1 Policy Improvement

This section deals with a very effective numerical method for solving stochastic optimal control problems for finite-state controlled Markov chains. The situation we consider is of a controlled continuous-time[3] Markov process $(X_t)_{t \geq 0}$ with values in the finite set I, and controlled by a control variable $a \in A$. At one stage of the argument (3.5), we need to be able to assume that certain suprema over $a \in A$ are finite and attained, which will certainly always happen if A is also finite. If the process is in state i, the jump intensity to state $j \neq i$ is $q_{ij}(a)$, a function of the control used; of course, the choice of control may depend on the state i, and we make the common

[3] The method works with the obvious changes also for discrete-time Markov chains.

notational convention for the diagonal entries of the intensity matrix

$$q_{ii}(a) \equiv -q_i(a) = -\sum_{j \neq i} q_{ij}(a).$$

The objective of the controller is to achieve

$$V(i) = \sup E\left[\int_0^\infty \exp\{-\int_0^t \rho(X_s, a_s)ds\}\, f(X_t, a_t)dt \,\middle|\, X_0 = i\right]. \tag{3.3}$$

Here, the functions $\rho : I \times A \to (0, \infty)$ and $f : I \times A \to \mathbb{R}^+$ are the discount rate, and the reward function respectively. Typically in the past we have supposed that ρ is constant, but there is no need to assume this.

We shall say that a map $\pi : I \to A$ is a *policy*; the interpretation is that $\pi(i)$ is the action that the policy would tell you to take when the state of the chain is i. The policy improvement algorithm is a recursive methodology, which starts with *any* policy $\pi^{(0)}$, and sequentially improves it by solving systems of linear equations. Here are the steps of the policy improvement algorithm to take us from policy $\pi^{(n)}$ to policy $\pi^{(n+1)}$:

1. Calculate and store the Q-matrix $Q^{(n)}$ defined by
$$Q_{ij}^{(n)} = Q_{ij}(\pi^{(n)}(i))$$
and functions $\rho^{(n)}(i) \equiv \rho(i, \pi^{(n)}(i))$, $f^{(n)}(i) \equiv f(i, \pi^{(n)}(i))$;
2. Solve the linear equation system
$$Q^{(n)}V^{(n)} - \rho^{(n)}V^{(n)} + f^{(n)} = 0 \tag{3.4}$$
for $V^{(n)}$;
3. For each i, calculate
$$\pi^{(n+1)}(i) = \text{argmax}_{a \in A}\Big\{\sum_j q_{ij}(a)V_j^{(n)} - \rho(i,a)V_i^{(n)} + f(i,a)\Big\}. \tag{3.5}$$

The working of the algorithm can be described very simply. Choose some policy $\pi^{(n)}$ and work out the value of using that policy; this is what $V^{(n)}$ is. Now check each state in turn, and see if there is an alternative action a available in that state which is 'better', in the sense that

$$\begin{aligned}
&\sum_j q_{ij}(a)V_j^{(n)} - \rho(i,a)V_i^{(n)} + f(i,a) \\
&\quad > \sum_j q_{ij}(\pi_i^{(n)})V_j^{(n)} - \rho(i, \pi_i^{(n)})V_i^{(n)} + f(i, \pi_i^{(n)}) = 0,
\end{aligned}$$

this last equality because of (3.4). What we will see in the proof is that if policy $\pi^{(n+1)}$ is better in this local sense than $\pi^{(n)}$, then it is also better globally.

To implement the algorithm, we have to be able to solve a system of linear equations, and to carry out a maximization[4] of some scalar function over the control variable. In applications, we find that not too many iterations are required, perhaps ten to twenty, so the whole algorithm runs fast. Just as importantly, as the following result shows, it runs stably. To avoid discussion of tiresome side issues, it is just stated under the assumption that A is finite.

Theorem 3.1 *Suppose that A is finite. The policy improvement algorithm generates a sequence of approximations to the value function V which increase monotonically:*

$$V^{(0)} \leq V^{(1)} \leq \ldots \leq V \tag{3.6}$$

and for some n large enough,

$$V^{(n)} = V. \tag{3.7}$$

Proof If π is a policy, define the operator $\mathscr{L}^{\pi}$ by

$$\begin{aligned}\mathscr{L}^{\pi} g(i) = E\Big[\int_0^{\tau_1} &\exp\Big(-\int_0^t \rho(X_s, a_s)\,ds\Big) f(X_t, a_t)\,dt \\ &+\exp\Big(-\int_0^{\tau_1} \rho(X_s, a_s)\,ds\Big) g(X_{\tau_1}) \,\Big|\, X_0 = i\Big],\end{aligned} \tag{3.8}$$

where τ_1 is the first time the chain changes state. Using basic properties of Markov chains, and writing $q_i \equiv -q_{ii}(\pi(i))$ we see that

$$\mathscr{L}^{\pi} g(i) = \frac{f(i, \pi(i))}{q_i + \rho(i, \pi(i))} + \frac{1}{q_i + \rho(i, \pi(i))} \sum_{j \neq i} q_{ij}(\pi(i)) g(j). \tag{3.9}$$

Now we know from (3.4) that $\mathscr{L}^{\pi^{(n)}} V^{(n)} = V^{(n)}$. If we write $\mathscr{L}^{\pi^{(n+1)}} V^{(n)}(i) \equiv V_i^{(n)} + \varepsilon_i$, and abbreviate $\pi_i^{(n+1)} = a_i, q_i \equiv -q_{ii}(\pi^{(n+1)})$, what (3.9) tells us is that

$$(q_i + \rho(i, a))(V_i^{(n)} + \varepsilon_i) = f(i, a) + \sum_{j \neq i} q_{ij}(a) V_j^{(n)}. \tag{3.10}$$

[4] In fact, exact maximization is not required in (3.5); finding some control which improves things will be enough provided we do not cycle round the algorithm not picking up enough gain. It is hard to state a sufficient set of conditions, but an inspection of the proof of Theorem 3.1 shows how the idea would work.

Rearranging this, and using (3.5) then (3.4), we learn that

$$\begin{aligned}(q_i + \rho(i,a))\varepsilon_i &= \sum_j q_{ij}(a)V_j^{(n)} - \rho(i,a)V_i^{(n)} + f(i,a)\\ &\geq \sum_j q_{ij}(\pi^{(n)})V_j^{(n)} - \rho(i,\pi^{(n)})V_i^{(n)} + f(i,\pi^{(n)})\\ &= 0. \end{aligned} \tag{3.11}$$

The conclusion is that $\varepsilon_i \geq 0$, and if there is a strict improvement at (3.5) then there will be strict inequality at (3.11), so ε_i will be strictly positive. The claim (3.6) is established (each $V^{(n)}$, being the value of *some* policy, is bounded above by the optimal V).

If the policy improvement algorithm is applied repeatedly, at any time when there is no change to the policy (3.5), then the value $V^{(n)}$ satisfies the dynamic programming equation so must be optimal. But in view of the finiteness of the set of policies being searched, a strict improvement can only be delivered finitely often, and eventually the algorithm terminates. □

3.1.1 Optimal Stopping

A special form of optimal control problem is the optimal stopping problem, where the only control available is either to continue letting the Markov chain evolve, or else to stop it. This problem can be cast into the form studied above, but is sufficiently commonly encountered that it is worth making explicit the form that the policy improvement algorithm takes for such a problem.

Suppose then that we are given a finite-state Markov chain with statespace I and jump intensity matrix Q. If we choose to stop when in state i we receive reward $r(i)$, and while we are in state i but not yet stopped we receive reward at rate $f(i)$. For simplicity of exposition, let us suppose that all rewards are discounted at constant rate $\rho > 0$, though as before this can be relaxed. The policy improvement algorithm sequentially constructs stopping sets $S^{(0)} = I \supseteq S^{(1)} \supseteq \ldots$ and corresponding value functions $V^{(0)} = r \leq V^{(1)} \leq \ldots$ which after finitely many steps of the algorithm will have found the optimal stopping rule, and which has the property that for each n

$$V_i^{(n)} = r_i \quad \forall i \in S^{(n)}. \tag{3.12}$$

Here is how the algorithm takes us from stopping set $S^{(n)}$ with corresponding value $V^{(n)}$ to stage $n+1$.

1. Calculate the vector $y \equiv QV^{(n)} - \rho V^{(n)} + f$, and then set

$$S^{(n+1)} = S^{(n)} \setminus \{i : y_i > 0\}; \tag{3.13}$$

2. Define $V^{(n+1)}$ by setting $V_i^{(n+1)} = r_i$ for all $i \in S^{(n+1)}$; by solving

$$0 = f_j - \rho x_j + \sum_{k \notin S^{(n+1)}} q_{jk} x_k + \sum_{k \in S^{(n+1)}} q_{jk} r_k \qquad (j \notin S^{(n+1)}); \qquad (3.14)$$

   and finally setting $V_j^{(n+1)} = x_j$ for $j \notin S^{(n+1)}$.

The reader should verify to his/her satisfaction that this algorithm achieves the optimal solution.

Remark An important use of the policy improvement algorithm just presented is in finding the least concave majorant (LCM) of some data $\{(x_i, y_i) : i = 1, \ldots, N\}$. Here we suppose that $x_1 < x_2 < \ldots < x_N$, and we want to find a sequence $(z_i)_{i=1}^N$ with the properties

1. $z_i \geq y_i$ for all i;
2. the piecewise linear function defined by $\{(x_i, z_i) : i = 1, \ldots, N\}$ is concave, so
$$\frac{z_i - z_{i-1}}{x_i - x_{i-1}} \quad \text{is decreasing in } i;$$
3. $(z_i)_{i=1}^N$ is minimal with properties 1 and 2.

If we define a Q-matrix by

$$\begin{aligned}
q_{i,i+1} &= (x_{i+1} - x_i)^{-1} \qquad (i = 2, \ldots, N-1);\\
q_{i+1,i} &= (x_{i+1} - x_i)^{-1} \qquad (i = 1, \ldots, N-2);\\
q_{1j} &= q_{Nj} = 0 \qquad (i = 1, \ldots, N)\\
q_{ij} &= 0 \quad \text{if } |i - j| > 1
\end{aligned}$$

with row sums zero, then the LCM is just the solution of the optimal stopping problem for this Markov chain with stopping reward function (y_i). Using the policy improvement algorithm is probably the fastest numerical method for calculating the LCM. It can be applied in higher dimensions also.

3.2 One-Dimensional Elliptic Problems

In most of the examples we studied in Chapter 2, time did not enter as an argument of the value function, so the HJB equation involved only derivatives with respect to one spatial variable. We intend to use policy improvement to solve such problems, but the first step is to spell out how we go about approximating a one-dimensional diffusion as a finite-state continuous-time Markov chain. Once we know how to do that, at each stage of the policy improvement algorithm we know how to set up the linear equations to be solved.

Suppose that we have a one-dimensional second-order linear ODE of the form

$$\mathscr{L} f(x) + h(x) = 0 \tag{3.15}$$

that we wish to solve in some interval $[a, b]$ with Dirichlet boundary conditions: $f(a) = y_a$, $f(b) = y_b$. The differential operator $\mathscr{L}$ is

$$\mathscr{L} = \tfrac{1}{2}\sigma(x)^2 \frac{d^2}{dx^2} + \mu(x) \frac{d}{dx} - \rho(x) \tag{3.16}$$

where ρ is non-negative, and we shall suppose that σ, ρ and μ are continuous (otherwise a discretization onto a grid has no hope of working).

We introduce a grid $I = \{a = x_0 < x_1 < \ldots < x_n = b\}$, not necessarily equally spaced, and intend to discretize the problem onto that grid. The operator (3.16) is the generator of a diffusion process, and we propose to approximate this by a Markov chain taking values in I. By the strong Markov property, it is enough to understand the behaviour of the process started at $x_i \in (a, b)$ and stopped at the stopping time $\tau = \inf\{t : X_t = x_{i+1} \text{ or } X_t = x_{i-1}\}$, because the complete path is made up of pieces of this form. Now the approximating Markov chain up to time τ is simple enough to describe; there is an intensity $\lambda_\pm$ of jumping from x_i to $x_{i\pm1}$. Thus for the Markov chain,

$$P^{x_i}[X_\tau = x_{i+1}] = \frac{\lambda_+}{\lambda_+ + \lambda_-}, \tag{3.17}$$

$$E^{x_i}[\tau] = \frac{1}{\lambda_+ + \lambda_-}. \tag{3.18}$$

Compare this with what happens to the diffusion. At each grid point $x_i \in (a, b)$, we approximate the diffusion by a Brownian motion with constant drift $\mu_i = \mu(x_i)$ and constant volatility $\sigma_i = \sigma(x_i)$. The exit properties of a Brownian motion with constant drift are well known. If we define $c = 2\mu_i/\sigma_i^2$, then the scale function s is characterized by $s'(x) = \exp(-cx)$, and so

$$\begin{aligned} P^{x_i}[X_\tau = x_{i+1}] &= \frac{s(x_i) - s(x_{i-1})}{s(x_{i+1}) - s(x_{i-1})} \\ &= \frac{e^{-cx_i} - e^{-cx_{i-1}}}{e^{-cx_{i+1}} - e^{-cx_{i-1}}} \\ &= \frac{e^{-cz-cx_i} - e^{-cz-cx_{i-1}}}{e^{-cz-cx_{i+1}} - e^{-cz-cx_{i-1}}} \end{aligned} \tag{3.19}$$

where $cz = \max\{-cx_{i-1}, -cx_{i+1}\}$. It is of course trivial that the last two are mathematically equivalent; they are not however numerically equivalent. The point is that we are certain that the arguments of all the exponentials in the final expression are *non-positive*, and so will present no problems numerically; the previous expression might involve differences of exponentials of large positive values, and this is numerical disaster.

Finally, if we want to compute the mean exit time, we set $g(x) = E^x[\tau]$ and observe that this must solve the ODE $\mathscr{L}g + 1 = 0$ with zero boundary conditions at the two ends of the interval. This is solved by

$$g(x) = Ae^{-cx} + B - x/\mu \tag{3.20}$$

where A and B are chosen to match the boundary conditions; we find that

$$A = \frac{x_{i+1} - x_{i-1}}{\mu(e^{-cx_{i+1}} - e^{-cx_{i-1}})}, \tag{3.21}$$

$$B = \frac{x_{i-1}e^{-cx_{i+1}} - x_{i+1}e^{-cx_{i-1}}}{\mu(e^{-cx_{i+1}} - e^{-cx_{i-1}})}. \tag{3.22}$$

From this, we calculate for the drifting Brownian motion the exit probability (3.19) and the mean exit time (3.20), and then choose $\lambda_{\pm}$ to make the two expressions (3.17), (3.18) match up. Of course, this calculation must be done for each grid point separately, since the values of μ_i and σ_i are local; nevertheless, this is not a problem.

One point to note is that the previous analysis assumed that the mean was non-zero, otherwise we are dealing with a martingale, and the expressions for the exit probabilities and mean exit times are given by limiting forms. This has to be coded as an exception case, since expressions such as (3.19) are numerically unstable for very small values of c.. The other exception which has to be dealt with is when the volatility σ_i is very small. In this case, the process locally looks like a constant drift μ, and this tells us what to do. If $\mu_i > 0$, then we just take $\lambda_+ = \mu_i/(x_{i+1} - x_i)$, and $\lambda_- = 0$.

This analysis tells us how we approximate the diffusion in the centre of the region, but what do we do at the endpoints? This needs some care, and is dealt with in Section 3.5; we postpone discussion till then, because the next topic is how we extend this approximation tale into higher dimensions.

3.3 Multi-Dimensional Elliptic Problems

The kinds of approximation techniques developed in Section 3.2 can serve again in higher dimensions, but there are some difficulties in applying them straight away. We shall explain these difficulties (and how to get round them) only in two dimensions, where simple plots illustrate what is going on. Once the reader has understood the two dimensional case, it will be obvious what to do in higher dimensions. We shall also suppose without any real loss of generality that we are going to try build a Markov chain approximation on a regularly-spaced grid $G = \mathbb{Z}^2$.

So we will suppose that we are given a diffusion in $\mathbb{R}^2$ with generator

$$\mathscr{G} \equiv \tfrac{1}{2} a_{ij}(x) D_i D_j + \mu_i(x) D_i \tag{3.23}$$

where $a(\cdot)$ is smooth, positive-definite symmetric 2×2, and $\mu(\cdot)$ is smooth, with values in $\mathbb{R}^2$, and of course $D_i \equiv \partial/\partial x_i$. The simplest situation to handle is that in which the matrix a is *diagonal*, for then the two components of the diffusion are independent, and each of them can be approximated by an independent Markov chain on $\mathbb{Z}$. At each grid point $(k, \ell) \in G$, we calculate intensities $\lambda_{\pm}^{(i)}, i = 1, 2$ as explained in Section 3.2, taking drift $\mu_i(k, \ell)$ and volatility $\sqrt{a_{ii}(k, \ell)}$ in the construction of $\lambda_{\pm}^{(i)}$ for $i = 1, 2$. Then the Markov chain at the central point (k, ℓ) in Fig. 3.1 will jump to $(k \pm 1, \ell)$ with intensities $\lambda_{\pm}^{(1)}$, and to $(k, \ell \pm 1)$ with intensities $\lambda_{\pm}^{(2)}$.

What is to be done if the diffusion matrix is *not* diagonal? The problem is best understood in an example where $\mu \equiv 0$ and $a(k, \ell)$ has two eigenvectors not aligned with the coordinate axes, with one of the eigenvalues extremely small. Then the diffusion is locally in effect a one dimensional Brownian motion moving along a line that is slanted, as in Fig. 3.2. If we have a Markov chain jump from (k, ℓ) to one of the four nearest neighbours $(k, \ell \pm 1)$, $(k \pm 1, \ell)$, then *it is impossible to achieve zero variance in the direction orthogonal to the slanting direction of diffusion.*

This is at first sight very hard to deal with, but if we realize that the generator is an operator applied to functions, and we suppose that the function values at the grid points should be thought of as being *interpolated* in some suitable fashion from the values at the grid points, a way forward emerges. By inscribing a circle into the convex hull of the nearest neighbours, as illustrated in Fig. 3.3, we can then identify eigenvectors of $a(k, \ell)$ (shown in Fig. 3.3 as the cross in the circle). Now those components of the diffusion *are* locally independent, and can be dealt with as we

Fig. 3.1 The grid in two dimensions

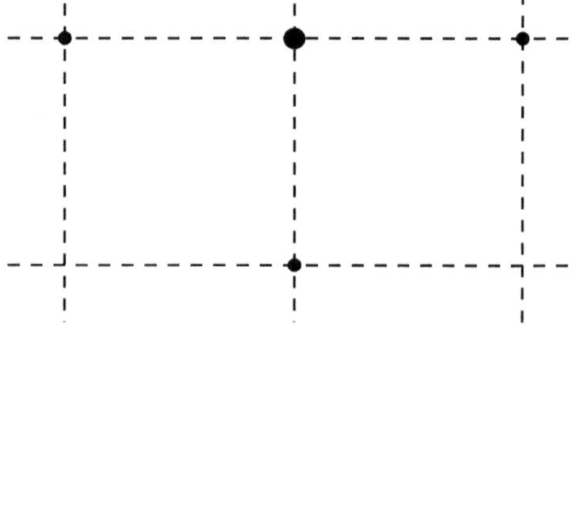

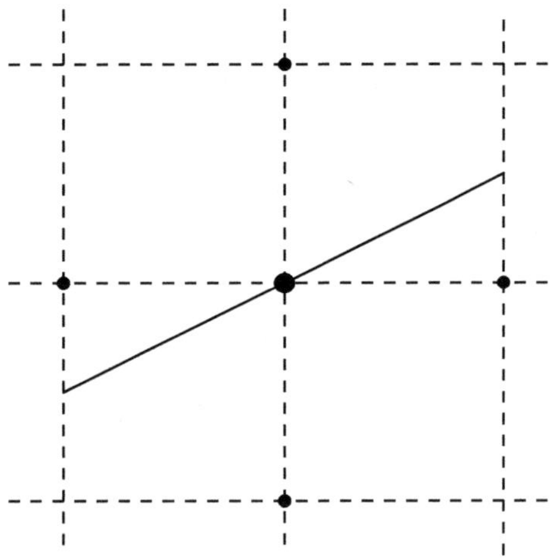

Fig. 3.2 A very elongated diffusion two dimensions

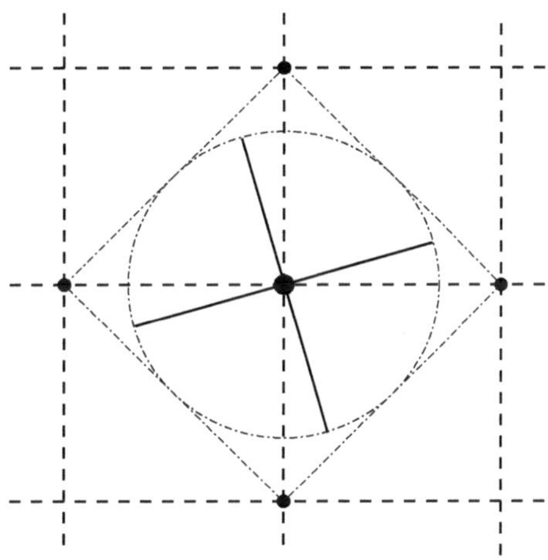

Fig. 3.3 How to interpolate function values to deal with a general diffusion

dealt with independent diffusions aligned with the coordinate axes. Doing so gives us four jump intensities from the central point (k, ℓ) to the ends of the cross. If the chain jumped with intensity α from the central point (now for simplicity assumed to be the origin) to (x_1, x_2) in the first orthant, the interpolated value of the function there would be just

$$f(x_1, x_2) = (1 - x_1 - x_2) f(0, 0) + x_1 f(1, 0) + x_2 f(0, 1). \tag{3.24}$$

Thus we ascribe an intensity αx_1 of jumping from the origin to $(1, 0)$ and an intensity αx_2 of jumping from the origin to $(0, 1)$. Other contributions come from the jumps to the other ends of the cross.

This is not the only possible way of thinking of the interpolation. We could instead think of jumping from the origin to the *corners* of the unit square, $(\pm 1, \pm 1)$, illustrated in Fig. 3.4 with the large black dots. This time, we would inscribe a circle inside the unit square, and identify the eigenvectors of $a(0,0)$ as shown in Fig. 3.5. If $(x_1, x_2) \in [-1, 1]^2$ is one of the ends of the cross, then we can see that as a convex combination

$$\begin{aligned}\begin{pmatrix} x_1 \\ x_2 \end{pmatrix} =& p_1 p_2 \begin{pmatrix} 1 \\ 1 \end{pmatrix} + p_1(1-p_2) \begin{pmatrix} 1 \\ -1 \end{pmatrix} \\ &+ (1-p_1)p_2 \begin{pmatrix} -1 \\ 1 \end{pmatrix} + (1-p_1)(1-p_2) \begin{pmatrix} -1 \\ -1 \end{pmatrix}, \end{aligned} \tag{3.25}$$

where $p_i = (1 + x_i)/2$, $i = 1, 2$. This is a smoother way of interpolating the values at the corners of the square, which may be preferable in some situations.[5] However, we cannot use the original grid $\mathbb{Z}^2$; we have to use $2\mathbb{Z}^2 \cup ((1,1) + 2\mathbb{Z}^2)$.

Fig. 3.4 Jumping to the corners of the square

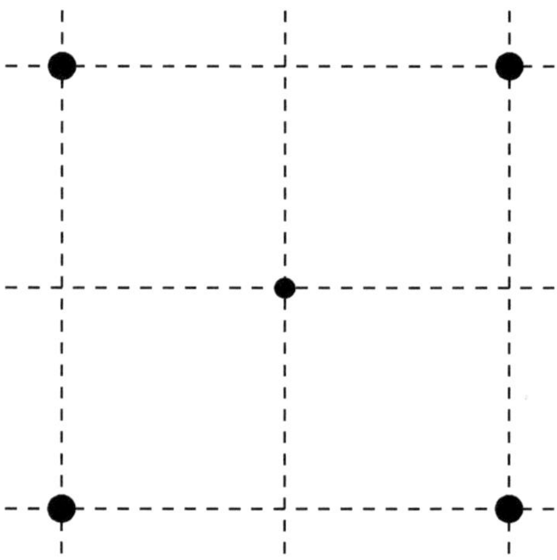

[5] Of course, in two dimensions, we could apply the second smooth interpolation recipe to the situation considered first, just by turning the picture through 45°, but this does not work in higher dimensions. The convex set we inscribed the circle into is the unit ℓ^1-ball in the first situation, with $2d$ vertices in d dimensions, whereas in the second situation we are working with the unit ℓ^∞-ball, with 2^d vertices.

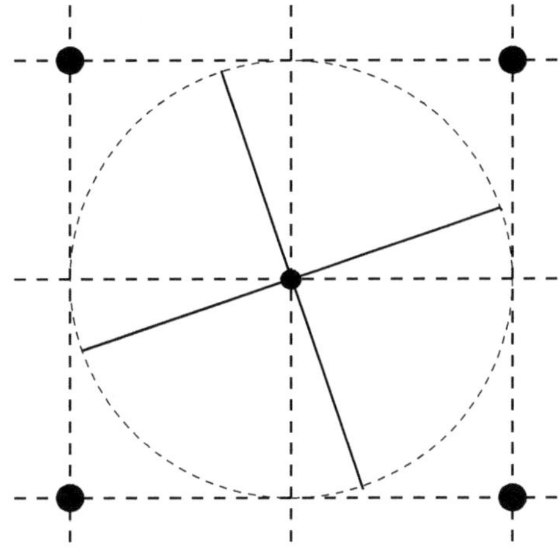

Fig. 3.5 The notional jumps of the Markov chain

3.4 Parabolic Problems

For simplicity of presentation, let us just discuss the case of a one-dimensional controlled diffusion[6]

$$dX_t = \sigma(X_t, a_t)\, dW_t + \mu(X_t, a_t)\, dt \tag{3.26}$$

where a_t is the control process, and the objective is

$$\sup E\Big[\int_0^T f(s, X_s, a_s)\, ds + F(X_T)\Big].$$

The value function

$$V(t, x) \equiv \sup E\Big[\int_t^T f(s, X_s, a_s)\, ds + F(X_T)\,\Big|\, X_t = x\Big] \tag{3.27}$$

solves the HJB equation

$$0 = \sup_a \big[\dot{V}(t, x) + \tfrac{1}{2}\sigma(x, a)^2 V''(t, x) + \mu(x, a) V'(t, x) + f(t, x, a)\big] \tag{3.28}$$

with the boundary condition $V(T, \cdot) = F(\cdot)$.

In order to solve this numerically, we shall suppose a grid $0 = t_0 < t_1 < \ldots < t_K = T$ of time points has been given, and that a grid $x_1 < \ldots < x_N$ of points on the line has been chosen. We shall make a finite-difference approximation to the

[6] We could allow the coefficients of the SDE to depend on time, but for notational simplicity we eschew this apparent generality.

differential operator $\mathscr{L}(a)$ defined on a smooth test function φ by

$$\mathscr{L}(a)(\varphi)(x) = \tfrac{1}{2}\sigma(x,a)^2\varphi''(x) + \mu(x,a)\varphi'(x). \tag{3.29}$$

Noticing that

$$\varphi'(x_i) \simeq \frac{\varphi(x_{i+1}) - \varphi(x_{i-1})}{\Delta_+ + \Delta_-} \tag{3.30}$$

$$\varphi''(x_i) \simeq \frac{\Delta_-\{\varphi(x_{i+1}) - \varphi(x_i)\} - \Delta_+\{\varphi(x_i) - \varphi(x_{i-1})\}}{\Delta_+\Delta_-(\Delta_+ + \Delta_-)} \tag{3.31}$$

where $\Delta_+ = x_{i+1} - x_i$, $\Delta_- = x_i - x_{i-1}$, it is possible to approximate the differential operator $\mathscr{L}(a)$ as a (sparse) tridiagonal matrix, which we shall denote by $L(a)$. If we abbreviate $\varphi(x_i) \equiv \varphi_i$, $\sigma_i = \sigma(x_i, a)$, $\mu_i = \mu(x_i, a)$, then the form of $L(a)$ is (for $1 < i < N$)

$$(L(a)\varphi)_i = q_i^+\,\varphi_{i+1} - q_i\,\varphi_i + q_i^-\,\varphi_{i-1}, \tag{3.32}$$

where

$$q_i^+ = \frac{\sigma_i^2}{2\Delta_+} + \mu_i$$

$$q_i = \tfrac{1}{2}\sigma_i^2\,(\Delta_+^{-1} + \Delta_-^{-1})$$

$$q_i^- = \frac{\sigma_i^2}{2\Delta_-} - \mu_i.$$

The analysis of Forsyth & Labahn [16] requires in its simplest form that

$$\text{Assumption } P: \quad q_i^{\pm} \text{are positive for all } i \text{ and for all } a,$$

and from now on we shall assume this condition.[7] It is clear that q_i will always be positive, and if the step sizes $\Delta_\pm$ are all quite small, the positive contribution $\sigma_i^2/2\Delta_\pm$ to $q_i^\pm$ will likely dominate the term $|\mu_i|$, so for a reasonably finely-spaced x-grid we may expect that Assumption P holds, though the requirement that positivity holds for all a can be problematic.

Now we discretize the partial differential operator appearing in the HJB equation (3.28) onto the chosen time and space grid, writing $V_i^n \equiv V(t_n, x_i)$, and $V^n = (V_i^n)_{i=1,\dots,N}$. What we obtain is

$$\frac{V^{n+1} - V^n}{t_{n+1} - t_n} + \theta\,\{L(a)V^n + f(t_n, \cdot, a)\} + (1-\theta)\{L(a)V^{n+1} + f(t_{n+1}, \cdot, a)\}, \tag{3.33}$$

[7] Quite possibly this condition is not needed; at the time of writing, this issue is not decided.

where $\theta \in [0, 1]$ determines the type of finite-difference scheme; $\theta = 0$ is the explicit scheme, $\theta = 1$ is the fully implicit scheme, and (most interesting for us) $\theta = \frac{1}{2}$ is the Crank-Nicolson scheme. For simplicity, we shall assume that the values V_1^n and V_N^n are fixed for every $n = 1, \ldots, K$, so that in effect there are absorbing boundary conditions at the ends of the x-range.

The way this discretized system of equations is going to be used is to do dynamic programming, starting at the final time $t_K = T$ and stepping backwards. The discretized form of (3.28) would tell us that

$$0 = \sup_a \left[\frac{V^{n+1} - V^n}{t_{n+1} - t_n} + \theta \{L(a)V^n + f(t_n, \cdot, a)\} \right.$$
$$\left. + (1-\theta)\{L(a)V^{n+1} + f(t_{n+1}, \cdot, a)\} \right]$$

for each $n = 0, \ldots, K-1$. We shall recursively determine the optimal controls a_i^n to be used at position x_i at time t_n by solving the optimization problems ($n = 0, \ldots, K-1$)

$$0 = \sup_a \left[\frac{V^{n+1} - V^n}{t_{n+1} - t_n} + \theta \{L(a)V^n + f(t_n, \cdot, a)\} \right.$$
$$\left. + (1-\theta)\{L(a^{n+1})V^{n+1} + f(t_{n+1}, \cdot, a^{n+1})\} \right]$$

and defining the optimal choice[8] of a to be a^n. This leaves just two questions to be answered:

- How do we start the recursion?
- How do we do the optimization?

To start the recursion, the natural thing to do is to take $\theta = 1$ to begin with when $n = K - 1$, so that we just solve a fully implicit scheme to get going, and we do not need to make any statement about what control we plan to use at the terminal time t_K. At all steps of the algorithm, because of Assumption P, *the optimization over a is the solution of an (elliptic) stochastic optimal control problem* which can therefore be achieved *using policy improvement!!* We can choose $\theta = \frac{1}{2}$ if we want to.

This approach of Forsyth and Labahn is a delightful combination of two of the slickest numerical methods—policy improvement and Crank-Nicolson—and is a very effective solution methodology for such problems.

---

[8] ... for simplicity assumed to exist and be unique ...

3.5 Boundary Conditions

As has already been said, getting the boundary conditions correct for a problem is usually critical to the effectiveness of the numerical scheme. The viewpoint of this chapter is that numerically solving a stochastic control problem is about exactly solving the analogous stochastic control problem for a finite-state Markov chain; and to do that, we have to tell a probabilistic story about how the Markov chain is going to behave when it gets to the boundary of the region.

The simplest story is to say that once the process hits the boundary, the process stops moving, and there is some terminal reward received. This was done in the habit formation example, Section 2.3, where we assumed at the lowest and highest grid point the value was given by assuming that the agent would come out of the risky asset completely, and receive the value he would get by consuming at constant rate (2.19) for ever. The probabilistic thinking behind this was that for very small values of wealth, the agent would have to keep his wealth volatility very small; and for very large values of wealth it would not really matter very much how he invested.

In some cases, such as the interest-rate risk example, Section 2.2, we imposed reflecting boundary conditions at the ends of a very large interval. While this is simple to do, it can be probabilistically rather implausible; if the state variable was wealth, for example, and the value were known to be an increasing function of wealth,[9] then for wealth values near to the lowest grid point we would not fear the wealth falling any lower, because the reflecting boundary assumption would cause the wealth to reflect back up. In such situations, all one can do is to push the boundary points to ever more extreme locations, and hope that (i) the numerics do not fail; (ii) in the middle of the region the solutions settle down. For the interest-rate risk example, reflecting boundaries at the ends are not stupid, because the state variable r is an OU process, and far away from its mean value the drift pushes strongly inwards in any case. The situation is rather analogous to the use of Fast Fourier Transform techniques for Fourier integrals; we know that any discretization of a Fourier integral will be a Fourier series, and therefore periodic, which is a property that the Fourier transform will not possess. Nevertheless, if we set the range of integration and the step size suitably, the Fourier series can be an excellent approximation to the Fourier integral over a desired range. The FFT methodology is just too powerful for this 'end effect' issue to make us abandon it, and the use of reflecting boundaries has a similar flavour.

But what else can we do? A better story would be to say that once the diffusion passes outside the region S where we are calculating the solution, some fixed controls are applied until the process re-enters S, when the controlled behaviour resumes. This is better than the first story, where we stop upon exit from S and receive the reward due if we used fixed controls from that time on, because we only lock down the controls while we are outside S, not for all time after we first exit S. Let us see how this works in one dimension.[10]

---

[9] The non-CRRA utility example of Section 2.8 is such an example.

[10] In higher dimensions, the story is much more complicated.

Suppose that the solution region is the interval $S = [\underline{x}, \bar{x}]$, and that outside S the controlled process evolves as a diffusion with generator

$$\mathscr{G} = \tfrac{1}{2}\sigma(x)^2 \frac{d^2}{dx^2} + \mu(x)\frac{d}{dx}. \tag{3.34}$$

Write ψ_ρ^+ (respectively, ψ_ρ^-) for the[11] non-negative increasing (respectively, decreasing) solution to

$$(\rho - \mathscr{G})\psi = 0,$$

so that the resolvent density may be expressed as[12]

$$r_\rho(x,y) = \frac{2}{\sigma(y)^2 W(y)}\psi_\rho^+(y\wedge x)\,\psi_\rho^-(y\vee x) \equiv h(y)\psi_\rho^+(y\wedge x)\,\psi_\rho^-(y\vee x), \tag{3.35}$$

where W is the Wronskian

$$W(x) = \psi_\rho^-(x)D\psi_\rho^+(x) - \psi_\rho^+(x)D\psi_\rho^-(x). \tag{3.36}$$

Consider now what would happen if the process started at some point $x > \bar{x}$, diffused with generator $\mathscr{G}$ until it hits $\bar{x}$, when it receives value $V(\bar{x})$. We would then be able to represent the value function at x as

$$V(x) = E^x\left[\int_0^\tau e^{-\rho t} f(X_t)\,dt\right] + E^x[e^{-\rho\tau}]\,V(\bar{x}), \tag{3.37}$$

where τ is the first passage time to $\bar{x}$. Rearranging this gives

$$\frac{V(x) - V(\bar{x})}{x - \bar{x}} = \frac{E^x\left[\int_0^\tau e^{-\rho t} f(X_t)\,dt\right] - E^x[1 - e^{-\rho\tau}]\,V(\bar{x})}{x - \bar{x}}. \tag{3.38}$$

The left-hand side of (3.38) tends to $V'(\bar{x})$ as $x \downarrow \bar{x}$, so we want to understand the limiting behaviour of the right-hand side. The resolvent density $r_\rho^\partial(x,y)$ of the diffusion killed on first hitting $\bar{x}$ is given for $\bar{x} < x < y$ by

$$r_\rho^\partial(x,y) = h(y)\psi_\rho^-(y)\{\psi_\rho^+(x) - \psi_\rho^+(\bar{x})\,\psi_\rho^-(x)/\psi_\rho^-(\bar{x})\}$$

and the derivative $q_\rho^+(y)$ of this with respect to x at $x = \bar{x}_+$ is easily seen to be

$$q_\rho^+(y) = h(y)\psi_\rho^-(y)\frac{W(\bar{x})}{\psi_\rho^-(\bar{x})} = \frac{2}{\sigma(y)^2}\frac{\psi_\rho^-(y)}{W(y)} \Big/ \frac{\psi_\rho^-(\bar{x})}{W(\bar{x})}. \tag{3.39}$$

---

[11] Unique up to scalar multiples...

[12] See Theorem V.50.7 in [34].

The boundary condition (3.38) therefore becomes in the limit as $x \downarrow \bar{x}$ the condition

$$V'(\bar{x}) = \int_{\bar{x}}^{\infty} q_\rho^+(y)\{f(y) - \rho V(\bar{x})\}dy, \tag{3.40}$$

which is a simple linear boundary condition of the form

$$AV(\bar{x}) + BV'(\bar{x}) + C = 0 \tag{3.41}$$

at the top boundary point $\bar{x}$. For the region below $\underline{x}$ we obtain analogously the derivative of the resolvent density is given by

$$q_\rho^-(y) = h(y)\psi_\rho^+(y)\frac{W(\bar{x})}{\psi_\rho^+(\bar{x})} = \frac{2}{\sigma(y)^2}\frac{\psi_\rho^+(y)}{W(y)}\Big/\frac{\psi_\rho^+(\bar{x})}{W(\bar{x})}, \tag{3.42}$$

leading to a boundary condition analogous to (3.40).

Notice that if we were to stop the process once it leaves S and give the agent a terminal reward, as we first discussed, this gives a boundary condition of the form (3.41) with $A = 1$, $B = 0$; if we reflect the diffusion at $\bar{x}$ we get a boundary condition of the form (3.41) with $A = 0 = C$, $B = 1$, so in every case we end up with a linear boundary condition of this form.

Example 1 If we take the generator to be

$$\mathscr{G} = \tfrac{1}{2}\sigma^2 D^2 + \mu D$$

for constant σ and μ, then

$$q_\rho^+(y) = \frac{2}{\sigma^2}\exp(-\alpha_+(y - \bar{x})) \qquad (y > \bar{x}) \tag{3.43}$$

$$q_\rho^-(y) = \frac{2}{\sigma^2}\exp(\alpha_-(y - \underline{x})) \qquad (y < \underline{x}) \tag{3.44}$$

where $\alpha_+ > 0$ and $-\alpha_- < 0$ are the roots of $\frac{1}{2}\sigma^2 t^2 + \mu t - \rho = 0$.

Example 2 If we take the generator to be

$$\mathscr{G} = \tfrac{1}{2}\sigma^2 x^2 D^2 + \mu x D$$

for constant σ and μ, then

$$q_\rho^+(y) = \frac{2}{\sigma^2 y^2}\left(\frac{y}{\bar{x}}\right)^{1-\alpha_+} \qquad (y > \bar{x}) \tag{3.45}$$

$$q_\rho^-(y) = \frac{2}{\sigma^2 y^2}\left(\frac{y}{\underline{x}}\right)^{1+\alpha_-} \qquad (y < \underline{x}) \tag{3.46}$$

where $\alpha_+ > 0$ and $-\alpha_- < 0$ are the roots of $\frac{1}{2}\sigma^2 t(t-1) + \mu t - \rho = 0$.

3.6 Iterative Solutions of PDEs

This short section is rather speculative, and very unfinished. All that is done here is to set down the form of the HJB equations quite generally, and then guess at possible recursive schemes which might be used to solve them; sometimes these quick-and-dirty schemes work, but I have no understanding at a theoretical level.

The form of the HJB equation for the value function V in an optimal investment problem can be written as

$$\mathscr{L}V + \sup_a \Psi(DV, a) = 0 \tag{3.47}$$

where $\Phi(\cdot, a)$ is some linear function of the vector[13] $DV \equiv (V, V', V'', \dot{V})$ of derivatives[14] of V, and a is the control variable. The operator $\mathscr{L}$ is a linear differential operator which does not depend on the control variable. Notice that the equation (3.47) states that a function of the state variable is zero; $\mathscr{L}V$ and $\Psi(DV, a)$ are each functions of the state variable.

Usually the optimization over control variable a can be achieved in closed form, and we may write

$$\sup_a \Psi(DV, a) = \bar{\Psi}(DV) \tag{3.48}$$

explicitly. The function $\bar{\Psi}$ is evidently convex in DV, and we shall suppose that it is C^1 in DV. When convenient in what follows, we shall suppose that the supremum in (3.48) is uniquely attained at the value $a^*(DV)$ of a.

The general form of the problem suggests possible recursive schemes.

3.6.1 Policy Improvement

As we have seen, probably the most effective general method is policy improvement. We begin by choosing some policy a_0, which is of course a function of the state variable, specifying what control is to be used in each state of the statespace. Then we set $n = 0$ and perform the iterative scheme:

(1) Find V_{n+1} by solving the *linear* system

$$\mathscr{L}V_{n+1} + \Psi(DV_{n+1}, a_n) = 0; \tag{3.49}$$

(2) Define the next choice of policy by

---

[13] We explain the method assuming that the state variable is one-dimensional, but the methodology works also in higher dimensions.

[14] For an infinite horizon problem, the value is not time dependent, so the time derivative $\dot{V}$ of V does not appear.

$$a_{n+1} = a^*(DV_{n+1}); \tag{3.50}$$

(3) Go to (1).

In the treatment of policy improvement in Section 3.1 we emphasized the probabilistic nature of the discretization used. However, just looking at the algorithm as it is listed above, it is tempting to just jump in and apply policy improvement using the finite-difference approximations to the differential operators, using natural boundary conditions. This has the drawback that now the linear system being solved does not necessarily have a probabilistic interpretation; but it has the advantages that it is easier to do, especially in higher dimensions.

3.6.2 Value Recursion

Another scheme we could try would be to suggest an initial guess V_0 for the value function, and then iteratively calculate V_n by

$$\mathscr{L}V_{n+1} + \bar{\Psi}(DV_n) = 0 \qquad (n = 0, 1, \ldots), \tag{3.51}$$

which is a sequence of *linear* equations. If we are lucky, the sequence of approximations V_n will converge, and when this happens, it is an easy coding exercise, and the computations run very fast.

Notice that this is *not* the same as the algorithm known as value improvement, which is an essentially discrete-time algorithm. Value improvement starts with a first guess $V^{(0)}$ at the value function, and then asks what would be the value we would obtain if we made one step and thereafter received $V^{(0)}$: recursively, we solve

$$V^{(n+1)}(x) = \sup_a \Big[f(x,a) + \beta E\{yV^{(n)}(X_1) \mid X_0 = x, a_0 = a\}y \Big], \quad n = 0, 1, \ldots. \tag{3.52}$$

under some boundedness conditions, the Contraction Mapping Principle applies, and we can deduce geometrically-rapid convergence of the $V^{(n)}$ to the true value function. As far as I am aware, there are no results telling us when and how rapidly value recursion converges to the true value function.

3.6.3 Newton's Method

Newton's method starts by making a guess for V_0, and then calculates successive approximations by writing $V_{n+1} = V_n + \eta$. We expect that η will be small if the method works at all, so the equation $\mathscr{L}V_{n+1} + \bar{\Psi}(DV_{n+1}) = 0$ may be approximated by a Taylor expansion of the non-linear term:

$$\mathscr{L}(V_n + \eta) + \bar{\Psi}(DV_n) + \nabla\bar{\Psi}(DV_n).D\eta = 0. \tag{3.53}$$

This then gives us a linear equation for η, which we can solve to find the next approximation V_{n+1}. This is the classical Newton method for finding a root of N simultaneous non-linear equations. It has the strengths and weaknesses of the Newton method: very rapid convergence if you are near to a solution, but small regions of convergence.

Chapter 4
How Well Does It Work?

Abstract The final chapter of the book takes a look at data, and finds virtually all of the models of the earlier part of the book to be wanting. Stylized facts of return data, well known to econometricians, are surprisingly robust across asset classes, and do not sit comfortably with the assumptions made in most of the theoretical literature.

For those who bought this book in the hope that it would help them to become rich, the short answer to the question of the title must be a disappointment: the answer is, 'Not very well at all'. There are two main reasons why this is the case:

(A) The actual dynamics of virtually any asset is not correctly modelled by the log-Brownian assumptions we have made;
(B) Even if the log-Brownian model were correct, it is hard to estimate the parameters of the model correctly.

We shall see in Section 4.1 below how simple exploratory data analysis conclusively rejects the standard log-Brownian model for an asset. We shall also see that it rejects most other commonly-used models for asset returns. Nevertheless, the situation is not as bleak as it seems. Most assets exhibit time-varying volatility on various scales; certainly within a day's trading there is a typical pattern where in the middle of the day the volatility is lower, which can be explained by a lunchtime effect, or more likely by the story that traders are particularly active at the beginning of a day when positions are adjusted for the coming day's trading, and at the end of the day when the aim is to leave the book in a safe state for the overnight period. If we look at daily prices, we observe that there are times when an asset may be quite heavily traded, and other times when there is less activity. A natural hypothesis is that there is some *business time* or *market clock*, an increasing process b_t, such that the observed asset price process S_t may be represented as

$$S_t = S^*(b_t) \tag{4.1}$$

L. C. G. Rogers, *Optimal Investment*, SpringerBriefs in Quantitative Finance,
DOI: 10.1007/978-3-642-35202-7_4, © Springer-Verlag Berlin Heidelberg 2013

where the process S^* is somehow more homogeneous and regular; this natural idea has been proposed in many places, the first of which appears to be Clark [6]. It turns out that this is actually quite a good story; if we scale daily asset returns by some estimate of current volatility, what we see *is plausibly stationary*, with a level of volatility which varies randomly but not too much. This is a very helpful observation; it means that we can rescale asset returns and then hope to estimate the more nearly time-homogeneous structure of those returns. From the point of view of the portfolio manager, it does not matter whether he specifies the position he is going to hold tomorrow in terms of the original asset, or in terms of the vol-rescaled asset, and the rescaled asset is much easier to understand statistically.

From the point of view of an investment bank selling and hedging derivatives, this observation is less useful, because although the underlying asset in business time may well look very like a log-Brownian model, the expiry of any option is not a fixed time in business time. So the writer of a derivative is exposed to the random fluctuation of the business time clock before expiry. Nonetheless, the understanding that the underlying asset is a log-Brownian motion sampled at a random time change is a powerful modelling insight, which converts the modelling problem into a problem of modelling the increasing business-time process. A number of popular and successful asset models, such as the variance-gamma model (see Madan and Seneta [26]), the CGMY and Meixner processes (see for example Madan and Yor [27] and references therein) have this structure.

So we see that the challenges posed by variable volatility are less of a problem to the portfolio manager than the derivatives business, which gives us a chance that Problem A will not matter so much here, and this is broadly correct. However, Problem B cannot be circumvented so easily, and there will be little advice here about how to deal with it, though there is some discussion of the nature and scale of the problem: see Sections 4.2 and 4.3. These econometric issues remain tough.

In the previous Chapters then, we followed the orthodoxy of mathematical research, and drew strong conclusions from strong assumptions; is there no value in any of this? If I believed this activity was valueless I would not have given so much of my time to it, but it is important to understand what the value is. In practice, we will never be able to say that the kinds of strong assumptions made[1] are valid, and even if we were so brave as to assume the form of asset dynamics, we would be foolish to assume that we knew the true parameter values exactly. But the examples studied earlier can give us *guidance* if not exact answers. For example, in the study of parameter uncertainty in Section 2.32, we saw that if we do not know the growth rate of the asset, then simply plugging the current posterior mean into the Merton proportion gives us an investment policy that is not far off optimal. If we care about the costs of trading, then the results of Section 2.4 tell us how we would modify the Merton trading/consumption recommendations, and we see that the portfolio bounds are remarkably wide (in fact, $O(\varepsilon^{1/3})$—see [32]),

[1] ... that asset dynamics are log-Brownian, with known parameters ...

and the impact of the transaction costs is remarkably small (in fact, $O(\varepsilon^{2/3})$—see Shreve and Soner [32, 38] again). This can already be helpful in formulating trading behaviour. If we want to control drawdown, then the results of Sections 2.5 and 2.30 at least suggest the form of the solution to use. Learning from our idealized studies, the relative magnitudes of the impact on efficiency of transaction costs, taxes, parameter uncertainty, infrequent portfolio review already help us to understand which market imperfections we need to worry about, and which are less important.

The remainder of this Chapter presents several topics. Firstly in Section 4.1 we take a look at some of the stylized facts of asset return data, and discover that market prices look very different from the log-Brownian story in fundamental ways. We follow this in Sections 4.2 and 4.3 by looking at some of the econometric issues around asset return data.

4.1 Stylized Facts About Asset Returns

Our working assumption through most of this book has been that assets are log Brownian motions, so in particular the daily log returns will be independent identically-distributed random variables. Is that in fact what the data shows? We shall take four large US stocks, the 3M Corporation (MMM), Alcoa Inc. (AA), Apple Inc. (AAPL), and American Electric Power Company Inc. (AEP) from 24th July 2000 to 19th July 2010.

Are returns actually Gaussian? The usual way to do this is to do a $q - q$ plot. How this works is you take the observed log returns X_i, $i = 1, \ldots, N$, and then re-order them to be increasing: $X_{(1)} \leq X_{(2)} \leq \ldots \leq X_{(N)}$. Now you calculate the quantiles $q_i = \Phi^{-1}((i - \frac{1}{2})/N)$, $i = 1, \ldots, N$ of the standard normal distribution function Φ, and you plot the pairs $(q_i, X_{(i)})$. If the X_i were normally distributed, the ith order statistic would be approximately $\mu + \sigma q_i$, so the plot would appear as a straight line. What we see is plotted in Fig. 4.1, for four large US stocks. Marked in the plots are vertical lines at -2 and 2; roughly 95 % of the data should lie between these. The straight line approximation may not be too bad in the middle of the region, but it is visibly not a convincing story across the whole range. We could do some more serious statistics at this point, but if you do some $q - q$ plots for many more assets you will be forced to accept that the Gaussian story does not fit the observed data.

Are returns stationary? The standard modelling assumption asserts that the returns are stationary. One nice diagnostic which explores this assumption is to plot the

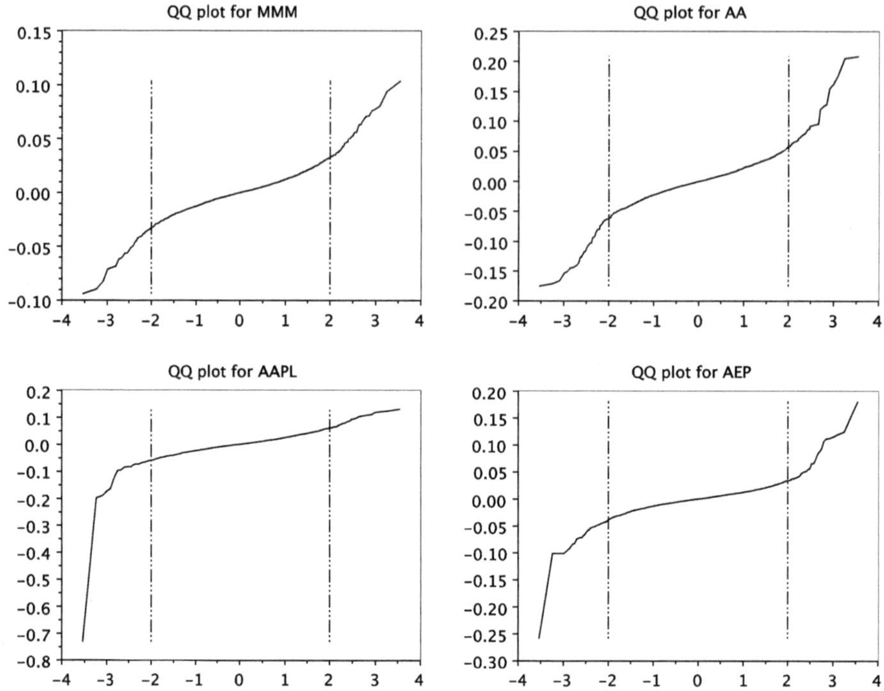

Fig. 4.1 $q - q$ plots of log returns for four US stocks

cumulative sum of squared returns; if the process were stationary[2] then we would expect the cumulative sums to go up linearly, and they evidently do not (Fig. 4.2).

Are returns autocorrelated? In view of the evidence against stationarity, you might (and should) be feeling uneasy about calculating the sample autocorrelation of asset returns, but please suspend disbelief a while, because the evidence continues to build. We calculate the sample autocorrelations of the returns of our four stocks in Fig. 4.3, and see that at positive lags, the numerical values are small; the dashed lines are at ± 0.1. Is there perhaps still some flicker of hope for the IID hypothesis? But no; Fig. 4.4 showing the autocorrelations of the *absolute* returns really extinguishes this—the autocorrelations remain substantially positive for a long time.

These little analyses are very simple to perform, and demonstrate well-known stylized facts of asset return data; see Granger et al. [17] for a more thorough study of what we find. You can amuse yourself with other such exploratory data analyses (and there are further examples at http://www.statslab.cam.ac.uk/~chris/Data.html,

[2] .. and square-integrable. Enthusiasts for heavy-tailed distributions may look at 4.2 and declare that this shows evidence for heavy-tailed returns—as they would when looking at 4.1. But if you simulate the cumulative sum of squared heavy-tailed random variables, it looks quite unlike what we see in 4.2; the big jumps in the simulations are quite clearly visible, whereas the plots from the data do not show any noticeable discontinuities.

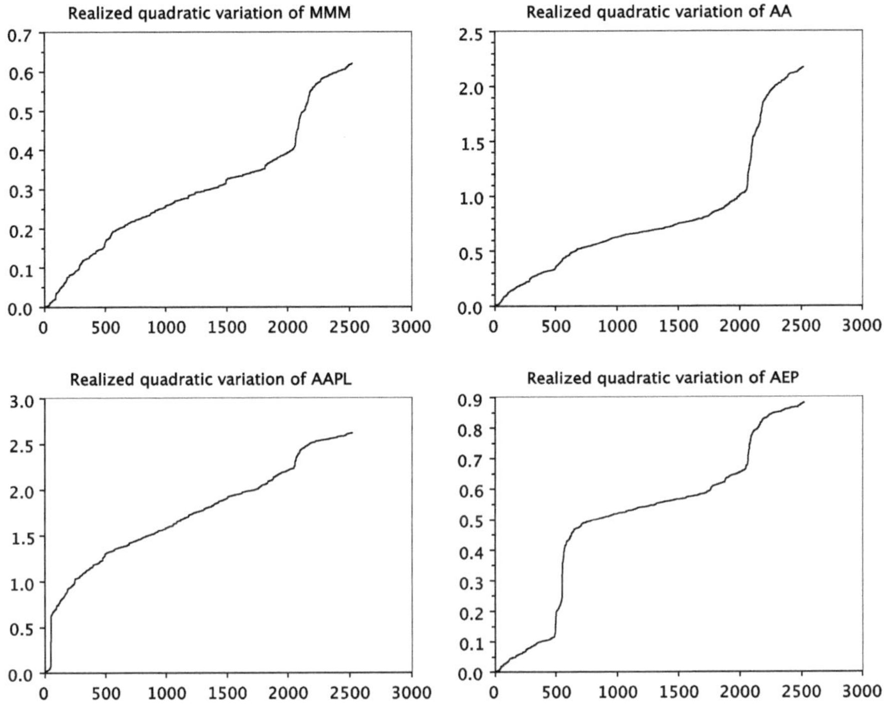

Fig. 4.2 Realized quadratic variation of four US stocks

but what they show is that the maintained assumption of IID Gaussian returns does not fit the facts. Even more, the assumption of IID returns of *any* law is seen not to fit the facts, which is a damaging observation for any attempt to model asset returns as log Lévy.

By contrast, if we firstly rescale the asset returns by a rolling estimate of the volatility[3] then we see somewhat different pictures. The $q - q$ plots are a little more regular, the realized quadratic variation is much closer to a straight line, as expected, the autocorrelation of returns is still small, and the autocorrelation of absolute returns is a bit smaller, but still quite some way from zero. So what we can conclude from this is that by vol-scaling asset returns we are able to produce returns which are reasonably stationary, which look a bit more Gaussian, have negligible autocorrelation, but still have persistent positive autocorrelation of absolute returns.

These stylized facts can be explained by some sort of *stochastic volatility model*; returns are zero mean, but the magnitudes of those returns do not stay constant, varying in some random fashion, with periods of large (or small) absolute returns persisting for some time. Really any stochastic volatility model would exhibit such behaviour; we could try some diffusion-based model, such as the Heston model (see

[3] In the plots that follow this was calculated as $\hat{\sigma}_t^2 = \sum_{j \geq 0} (1 - \beta)\beta^j \, r_{t-j}^2$ with $\beta = 0.975$, and $r_t \equiv \log(p_t / p_{t-1})$, where p_t is the close price on day t.

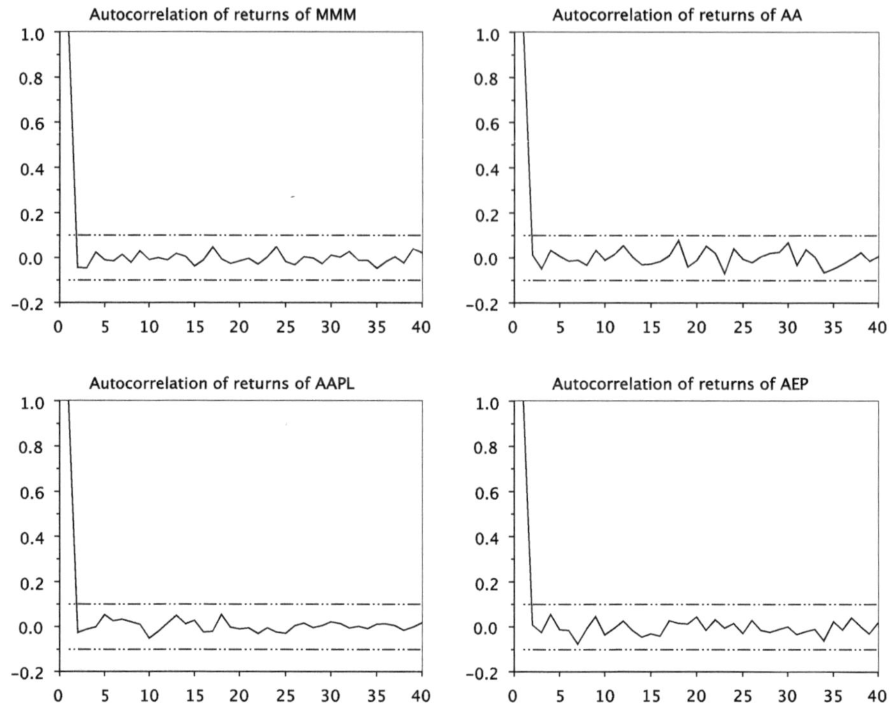

Fig. 4.3 Sample autocorrelation of four US stocks

[19]), we could try a GARCH model, popular with econometricians, we could take some Markov chain regime-switching model, for example.

For those unfamiliar with GARCH modelling, the basic GARCH(1,1) model for a process x_t in discrete time is defined by the recursive recipe

$$\begin{aligned} x_{t+1} &= x_t + v_{t+1}\varepsilon_{t+1} \\ v_{t+1} &= \alpha_0 + \alpha_1 v_t + \beta\varepsilon_t^2 \end{aligned}$$

for positive α_0, β and $\alpha_1 \in (0, 1)$. The ε_t are usually taken to be IID standard Gaussians, and of course, some starting values have to be given. Although the GARCH model is well established, I find it an unattractive modelling choice for a number of reasons (Figs. 4.5, 4.6, 4.7 and 4.8):

1. It is a discrete-time model which cannot be embedded into any continuous-time model; that is, there is no time-homogeneous continuous-time process[4] X_t which, when viewed at integer times, is a GARCH process;

---

[4] Of course, we could just have a continuous time process which jumps only at integer times, but this would not be time homogeneous.

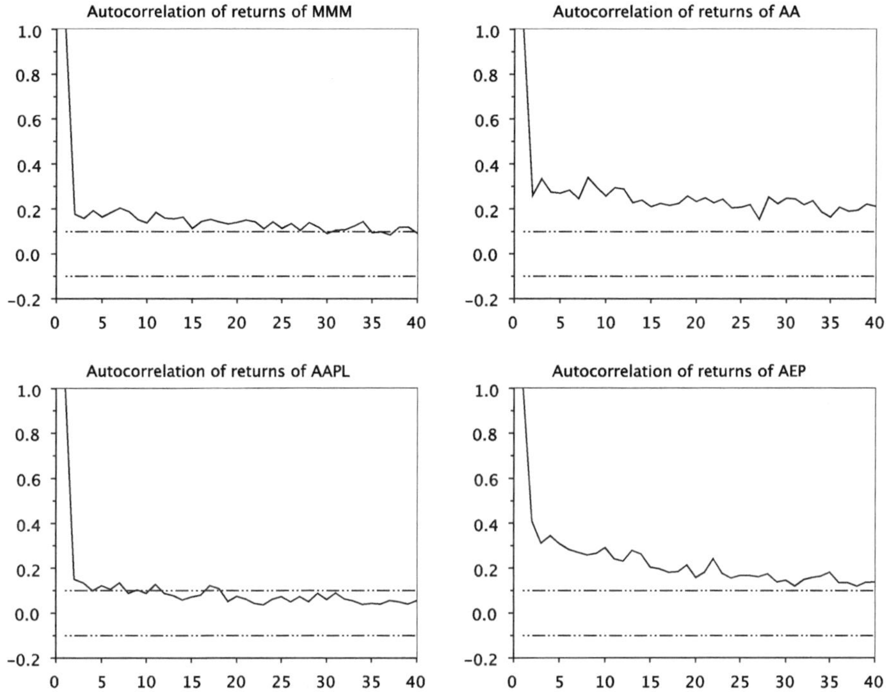

Fig. 4.4 Sample autocorrelation of the absolute returns of four US stocks

2. Aggregation does not work for a GARCH process; if X_n is a GARCH process, then X_{2n} is nothing in particular;
3. Multiple assets do not fit well into the framework. If we assume (reasonably) that we need to be able to model co-dependence of different GARCH series, how is this to be done naturally? In the basic GARCH story, the process generates its own volatility, yet a stylized fact of asset returns is that *they all experience high volatility at the same time.* The plot Fig. 4.9 shows what happens when we plot exponentially-weighted moving averages of squared daily returns for 29 US stocks. We could perhaps take yesterday's squared returns of all the assets, and use an average value of this to provide the increment for the volatility of each of the assets, but this kind of thinking is taking us towards modelling a market clock, and if that is where we are going, we would probably have been better not to start with GARCH.

To conclude our brief scan of market data then, we see that the paradigm model used extensively throughout this book fails significantly to match stylized facts. This can be rescued to some extent by working with volatility-rescaled returns, but some kind of stochastic volatility model is required to do a decent job on the stylized facts. The examples from Sections 2.10 and 2.26 are the only ones we have studied with this character.

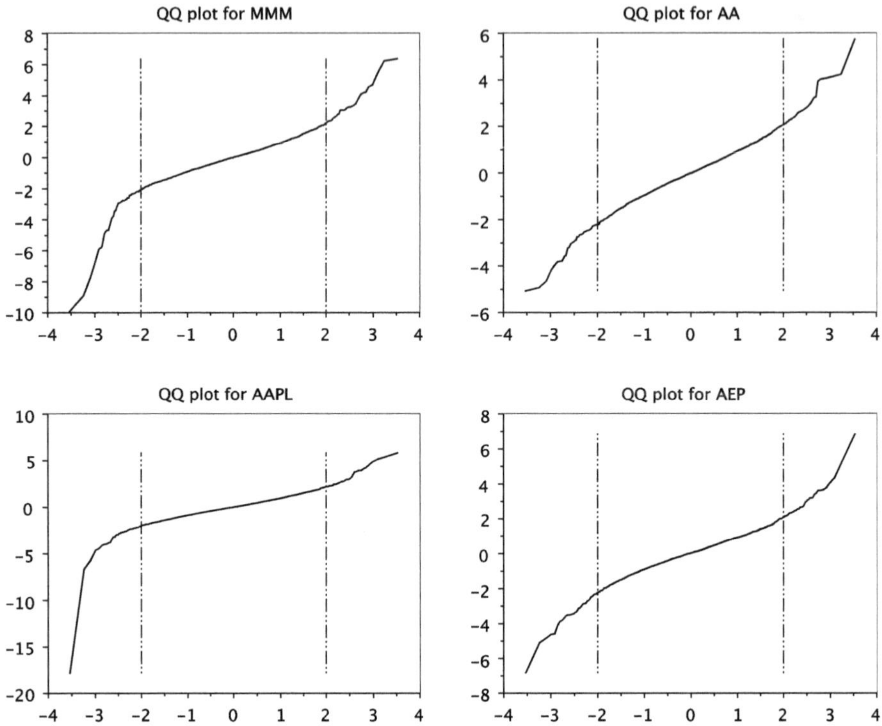

Fig. 4.5 $q-q$ plots of scaled log returns for four US stocks

4.2 Estimation of μ: The 20s Example

This little example, which requires no more than an understanding of basic statistical concepts, should be remembered by anyone who works in finance. It is memorable because all the numbers appearing are something to do with 20.

Suppose we consider a stock, with annualised rate of return $\mu = 0.2 = 20\,\%$, and annualised volatility $\sigma = 0.2 = 20\,\%$. We see daily prices for N years, and we want to observe for long enough that our 95 % confidence interval for σ (respectively, μ) is of the form $[\hat{\sigma} - 0.01, \hat{\sigma} + 0.01]$ (respectively, $[\hat{\mu} - 0.01, \hat{\mu} + 0.01]$)—so we have a 19 in 20 chance of knowing the true value to one part in 20.

How big must N be to achieve this precision in $\hat{\sigma}$*?*

Answer: about 13 years;

How big must N be to achieve this precision in $\hat{\mu}$*?*

ANSWER: about 1580 years !!

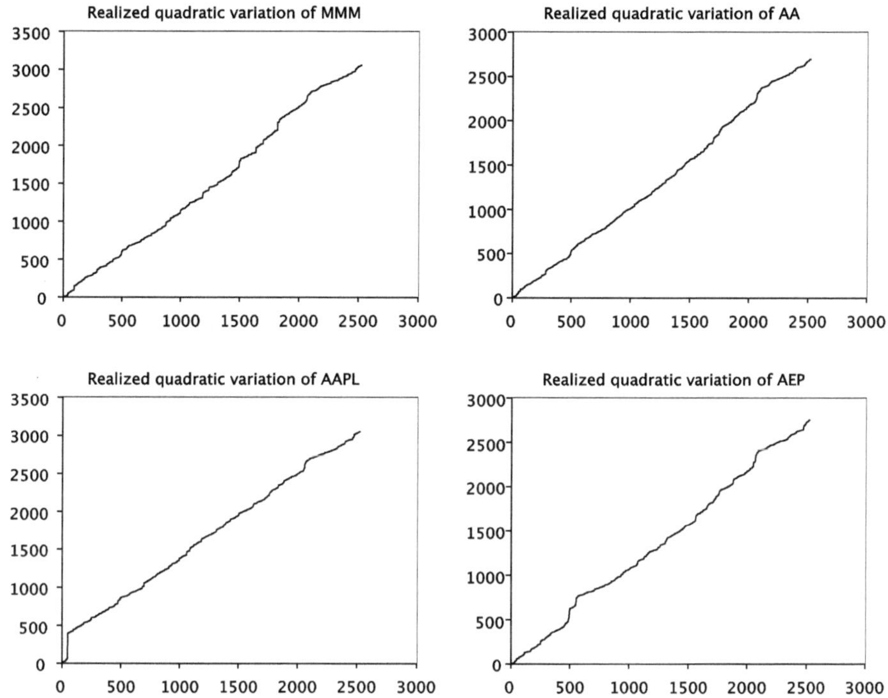

Fig. 4.6 Realized (scaled) quadratic variation of four US stocks

The message here is that the volatility in a typical financial asset overwhelms the drift to such an extent that we cannot hope to form reliable estimates of the drift without centuries of data. This underscores the pointlessness of trying to fit some model which tells a complicated story about the drift; if we cannot even fit a constant reliably, what hope is there for fitting a more complicated model?

Could we improve our estimates if we were to observe the asset price more frequently, perhaps every hour, or every minute? In principle, by doing this we could estimate σ to arbitrary precision, because the quadratic variation of a continuous semimartingale is recoverable path by path. But there are practical problems here. Most assets do not fluctuate at the constant speed postulated by the simple log-Brownian model, and the departures from this are more evident the finer the timescale one observes[5]; thus we will not arrive at a certain estimate just by observing the price every 10 s, say. The situation for estimation of the drift is even more emphatic; since

[5] In recent years, there has been an upsurge in the study of realized variance of asset prices; an early reference is Barndorff-Nielsen and Shephard [2], a more recent survey is Shephard [37], and there have been important contributions from Aït-Sahalia, Jacod, Mykland, Zhang and many others. This literature is concerned with estimating what the quadratic variation actually was over some time period, which helps in deciding whether the asset price process has jumps, for example. However, there is no parametric model being fitted in these studies; the methodology does not claim or possess predictive power.

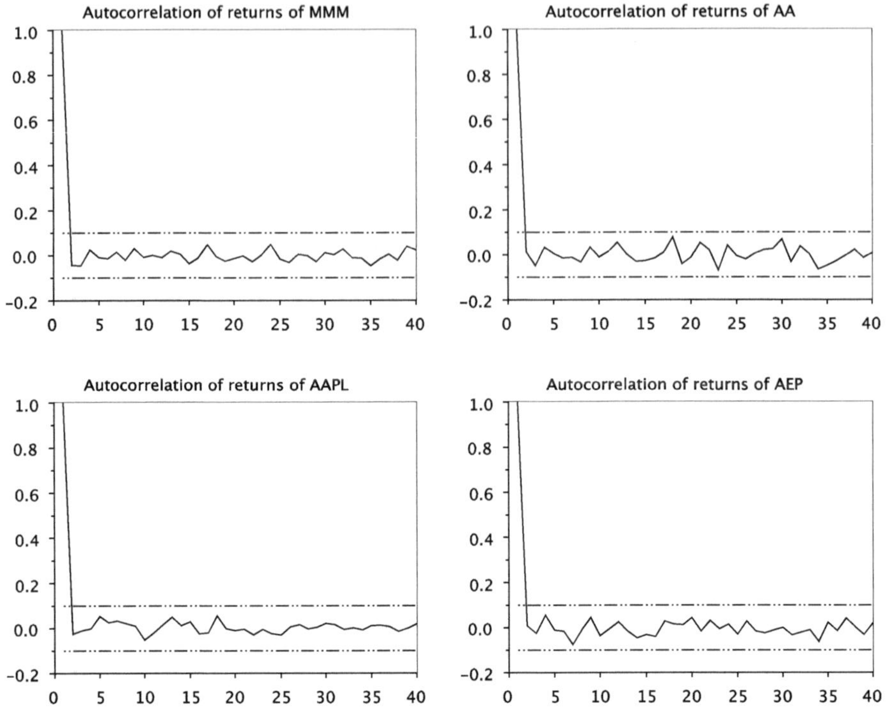

Fig. 4.7 Sample autocorrelation of scaled returns of four US stocks

the sum of an independent gaussian sample is sufficient for the mean, observing the prices more frequently will not help in any way to improve the precision of the estimate of μ; the change in price over the entire observation interval is the only statistic that carries information about μ.

The most important thing to know about the growth rate of a financial asset is that you don't know it.

4.3 Estimation of V

The conclusion of the 20 s example suggests that the estimation of σ is less problematic than the estimation of μ; we may be able to form a decent estimate of σ in a decade or so, perhaps less if we sample hourly during the trading day. However, the situation is not as neat as it appears. Firstly, the assumption of constant σ is soundly rejected by the data; this, after all, was a major impetus for the development of GARCH models of asset prices. Secondly, and just as importantly, the estimation of σ in *multivariate* data is fraught with difficulty. To show some of the issues, suppose that we observe daily log-return data $X_1, \ldots, X_T$ on N assets, where $X_t = (X_t^i)_{i=1}^N$,

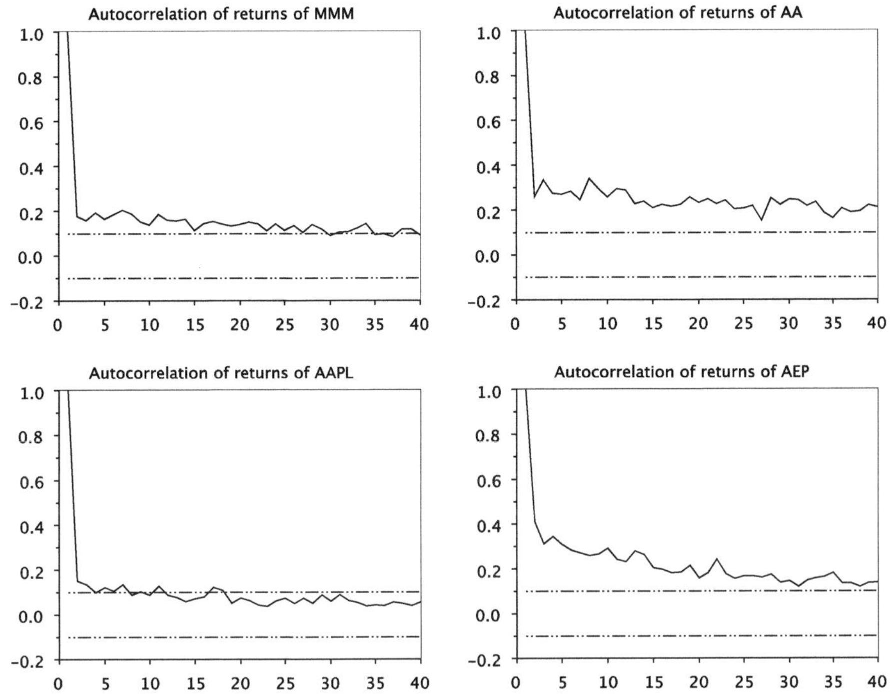

Fig. 4.8 Sample autocorrelation of the absolute scaled returns of four US stocks

$X_t^i = \log(S_t^i/S_{t-1}^i)$. The canonical maximum-likelihood estimator of the mean of X_t is to use

$$\hat{\mu} = T^{-1} \sum_{t=1}^{T} X_t$$

and to estimate the variance we use the sample covariance

$$\hat{V} = T^{-1} \sum_{t=1}^{T} (X_t - \hat{\mu})(X_t - \hat{\mu})^T. \tag{4.2}$$

Just to get an idea, we display in Fig. 4.10 the correlations between some 29 US stocks; as can be seen, correlations are generally positive, and range widely in value from 0 to around 0.7. Such behaviour is quite typical.

But what are the snags?

1. For $N = 50$, there are 1275 independent parameters to be estimated in V;
2. The estimator of V is not very precise (if $T = 1000$, and $N = 50$, from simulations we find that the eigenvalues of $\hat{V}$ typically range from 0.6 to 1.4, while the true values are of course all 1.)

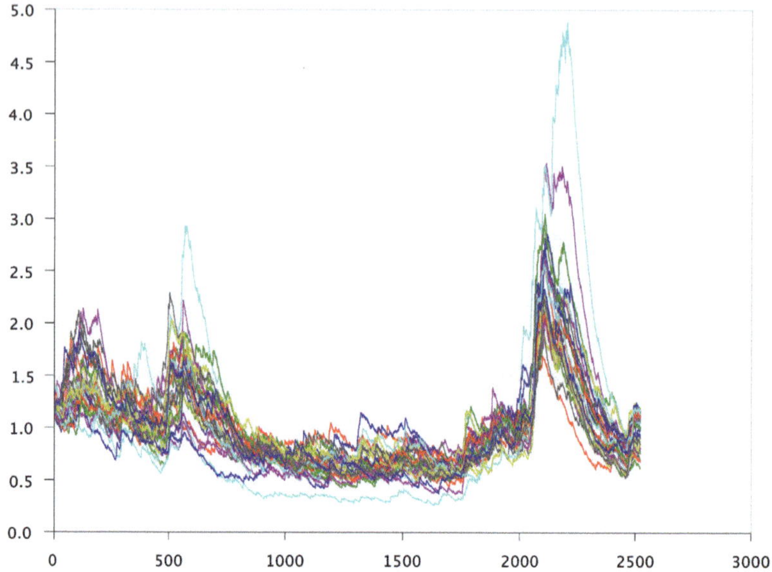

Fig. 4.9 Rolling volatility estimates of 29 US stocks

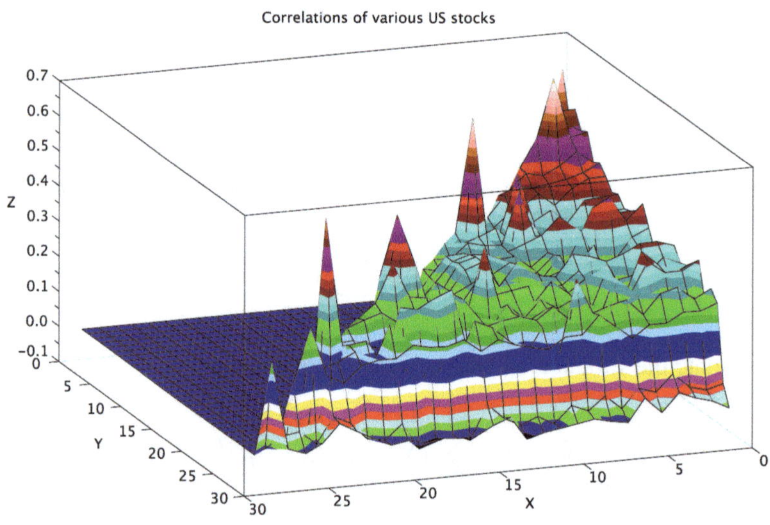

Fig. 4.10 Correlation plot

3. To form the Merton portfolio, we must invert V; inverting $\hat{V}$ frequently leads to absurdly large portfolio values.

All of these matter, but perhaps the first matters most; the number of parameters to be estimated will grow like N^2, and for N of the order of a few score—not

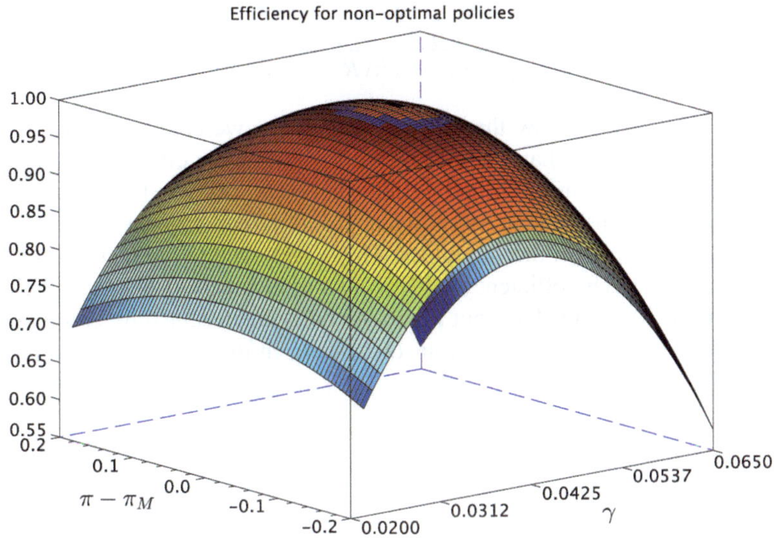

Fig. 4.11 Efficiency plot

an unrealistic situation—we have thousands of parameters to estimate. The only reasonable way to proceed is to cut down the dimension of the problem. One way in which to do this would be to insist that the correlations between assets were constant. This is a pretty gross assumption. Another thing one could do would be to perform a principal-components analysis, which in effect would just keep the top few eigenvalues from the spectrum of $\hat{V}$, which in any case account for most of the trace in typical examples. Yet another approach would be to suppose that the asset returns are linear combinations of the returns on a fairly small set of economic indicators which are considered important. All of these approaches are used in practice, and the literature is too large to survey here; one could begin with Fan and Lv [15] or Fan, Liao and Mincheva [14], for example.

How sensitive is the value of the Merton problem to the choice of the portfolio proportions and the consumption rate? If the agent chooses a consumption rate γ, and to keep proportions $\pi = \pi_M + \varepsilon$ of his wealth in the risky assets, then we can use (1.78), expressing the value of the objective as

$$\frac{u(\gamma w_0)}{R(\gamma_M - \gamma) + \gamma - \frac{1}{2}R(R-1)|\sigma^T \varepsilon|^2}, \tag{4.3}$$

which we see reduces to the Merton value $\gamma_M^{-R} u(w_0)$ when $\gamma = \gamma_M$ and $\varepsilon = 0$. This allows us to find the efficiency of an investor who uses sub-optimal policy $(\gamma, \pi_M + \varepsilon)$, namely,

$$\theta = \left(\frac{\gamma_M^R \gamma^{1-R}}{R(\gamma_M - \gamma) + \gamma - \frac{1}{2}R(R-1)|\sigma^T \varepsilon|^2} \right)^{1/(1-R)}. \tag{4.4}$$

The plot Fig. 4.11 shows how the efficiency varies as we change γ and π, using the default values (2.3), What is most noteworthy is that the efficiency is not much affected by the wrong choice of π and γ. Indeed, we can vary γ in the interval (0.033, 0.053) without losing more than 5 % efficiency, and we can vary the proportion π in the range (0.22, 0.52) with the same loss. This is very robust, though on reflection not a great surprise. The efficiency will be a smooth function of (γ, π), which is maximized at the Merton values, but it will of course have vanishing gradient there, and so the variation in efficiency for an $O(h)$ error in the choice of (γ, π) will be $O(h^2)$.

References

1. T. Arun, The Merton problem with a drawdown constraint on consumption. arXiv 1210:5205v1 (2012)
2. O.E. Barndorff-Nielsen, N. Shephard, Econometric analysis of realized volatility and its use in estimating stochastic volatility models. J. R. Stat. Soc. Ser. B **64**, 253–280 (2002)
3. T. Björk, A. Murgoci, A general theory of Markovian time inconsistent stochastic control problems. Technical report, Department of Finance, Stockholm School of Economics, 2010
4. K.C. Border, *Fixed Point Theorems with Applications to Economics and Game Theory* (Cambridge University Press, Cambridge, 1985)
5. D.T. Breeden, Consumption, production, inflation and interest rates: a synthesis. J. Financ. Econ. **16**, 3–39 (1986)
6. P.K. Clark, A subordinate stochastic process model with finite variance for speculative prices. Econometrica **41**, 135–155 (1973)
7. G.M. Constantinides, Capital market equilibrium with transaction costs. J. Politi. Econ. **94**, 842–862 (1986)
8. G.M. Constantinides, Habit formation: a resolution of the equity premium puzzle. J. Politi. Econ. **98**, 519–543 (1990)
9. T.M. Cover, Universal portfolios. Math. Financ. **1**, 1–29 (1991)
10. M.H.A. Davis, A. Norman, Portfolio selection with transaction costs. Math. Oper. Res. **15**, 676–713 (1990)
11. I. Ekeland, A. Lazrak, Being serious about non-commitment: subgame perfect equilibrium in continuous time. Technical report, University of British Columbia, 2006
12. N. El Karoui, S. Peng, M.C. Quenez, Backward stochastic differential equations in finance. Math. Financ. **7**, 1–71 (1997)
13. R. Elie, N. Touzi, Optimal lifetime consumption and investment under a drawdown constraint. Financ. Stoch. **12**, 299–330 (2006)
14. J. Fan, Y. Liao, M. Mincheva, High dimensional covariance matrix estimation in approximate factor models. Ann. Stat. **39**, 3320–3356 (2011)
15. J. Fan, J. Lv, A selective overview of variable selection in high dimensional feature space. Stat. Sin. 20, 101–148 (2010)
16. P.A. Forsyth, G. Labahn, Numerical methods for controlled Hamilton-Jacobi-Bellman PDEs in finance. J. Comput. Financ. **11**, 1–43 (2007)
17. C.W.J. Granger, S. Spear, Z. Ding, Stylized facts on the temporal and distributional properties of absolute returns: an update. in *Statistics and Finance: An Interface. Proceedings of the Hong Kong International Workshop on Statistics in Finance*, 2000
18. D. Heath, R. Jarrow, A. Morton, Bond pricing and the term structure of interest rates: a new methodology for contingent claims valuation. Econometrica **60**, 77–105 (1992)

L. C. G. Rogers, *Optimal Investment*, SpringerBriefs in Quantitative Finance,
DOI: 10.1007/978-3-642-35202-7, © Springer-Verlag Berlin Heidelberg 2013

19. S.L. Heston, A closed-form solution for options with stochastic volatility with applications to bond and currency options. Rev. Financ. Stud. **6**, 327–343 (1993)
20. D.G. Hobson, L.C.G. Rogers, Complete models of stochastic volatility. Math. Financ. **8**, 27–48 (1998)
21. I. Karatzas, S.E. Shreve, *Methods of Mathematical Finance* (Springer, New York, 1998)
22. I. Klein, L.C.G. Rogers, Duality in constrained optimal investment and consumption problems: a synopsis. Math. Financ. **17**, 225–247 (2007)
23. D. Kramkov, W. Schachermayer, The asymptotic elasticity of utility functions and optimal investment in incomplete markets. Ann. Appl. Probab. **9**, 904–950 (1999)
24. H.J. Kushner, P.G. Dupuis, *Numerical Methods for Stochastic Control Problems in Continuous Time* (Springer, New York, 2000)
25. B.H. Lim, Y.H. Shin, Optimal investment, consumption and retirement decision with disutility and borrowing constraints. Quant. Financ. **11**, 1581–1592 (2011)
26. D.B. Madan, E. Seneta, The variance gamma (VG) model for share market returns. J. Bus. **63**, 511–524 (1990)
27. D.B. Madan, M. Yor, Representing the CGMY and Meixner Lévy processes as time changed Brownian motions. J. Comput. Financ. **12**, 27–48 (2008)
28. R.C. Merton, Optimum consumption and portfolio rules in a continuous-time model. J. Econ. Theory **3**, 373–413 (1971)
29. R. Muraviev, L.C.G. Rogers, Utilities bounded below. Ann. Finance
30. B. Øksendal, *Stochastic Differential Equations: An Introduction with Applications*, 6th edn. (Springer, Berlin, 2012)
31. L.C.G. Rogers, Duality in constrained optimal investment and consumption problems: a synthesis. in *Paris-Princeton Lectures on Mathematical Finance, vol. 1814 of Lecture Notes in Mathematics* (Springer, Berlin, 2003), pp. 95–131
32. L.C.G. Rogers, Why is the effect of proportional transaction costs $O(\delta^{2/3})$? in *Mathematics of Finance: Proceedings of an AMS-IMS-SIAM Joint Summer Research Conference on Mathematics of Finance, 22–26 June 2003, Snowbird, Utah, Contemporary Mathematics*, vol. 351, ed. by G. Yin, Q. Zhang (American Mathematical Society, Providence, 2004), pp. 303–308
33. L.C.G. Rogers, D. Williams, *Diffusions, Markov Processes, and Martingales*, vol. 1, 2nd edn. (Cambridge University Press, Cambridge, 2000)
34. L.C.G. Rogers, D. Williams, *Diffusions, Markov Processes, and Martingales*, vol. 2, 2nd edn. (Cambridge University Press, Cambridge, 2000)
35. L.C.G. Rogers, M.R. Tehranchi, Can the implied volatility surface move by parallel shifts? Financ. Stoch. **14**, 235–248 (2010)
36. D. Romer, *Advanced Macroeconomics*, 2nd edn. (McGraw Hill, New York, 2001)
37. N. Shephard, *Stochastic Volatility: Selected Readings* (Oxford University Press, Oxford, 2005)
38. S.E. Shreve, H.M. Soner, Optimal investment and consumption with transaction costs. Ann. Appl. Probab. **4**, 609–692 (1994)

Index

20's example, 144

A
Admissible, 2
Advisors, 108
Annual tax accounting, 43
Asset price process, 2
Autocorrelation, 140

B
Bankruptcy, 50
Bayesian analysis, 108
Beating a benchmark, 94
Benchmark, 75
Black–Scholes–Merton model, 28
Boundary condition, 130
  linear, 132
  reflecting, 130
Brownian integral representation, 14, 16
BSDE, 72
Budget constraint, 94
Budget feasible, 14
Business time, 137

C
Cameron–Martin–Girsanov theorem, 103
Central planner, 25
Change of measure martingale, 15
Complete market, 14, 25
Compound Poisson process, 50
Consumption stream, 2
Contraction mapping principle, 135
Crank–Nicolson scheme, 117, 129
CRRA utility, 6

D
Default parameter values, 29
Depreciation, 81
Drawdown constraint
  on wealth, 39
Drawdown constraint
    on consumption, 64
Dual feasibility, 19
Dual HJB equation, 19
  drawdown constraint on consumption, 66
  drawdown constraint on wealth, 40
  habit formation, 35
  labour income, 111
  stopping early, 69
Dual objective, 19
Dual value function, 11, 48
  retirement, 101

E
Efficiency, 28
Elliptic problem, 121
  multi-dimensional, 123
Endowment, 2
Equilibrium, 23, 76
Equilibrium interest rate, 24
Equilibrium price, 27
Equivalent martingale measure, 19
Estimation of *V*, 146
Expected shortfall, 170

F
Fast Fourier Transform, 149
Filtering, 102
Financial review, 79
Finite-horizon Merton problem, 30

L. C. G. Rogers, *Optimal Investment*, SpringerBriefs in Quantitative Finance,
DOI: 10.1007/978-3-642-35202-7, © Springer-Verlag Berlin Heidelberg 2013

F (*cont.*)
Fixed-mix rule, 108
Forward rates, 23

G
GARCH model, 142

H
Habit formation, 33
Hamilton–Jacobi–Bellman. *See* HJB equation
Heat equation, 116
Heston model, 141
History-dependent preferences, 45
HJB equation, 5
  drawdown constraint
    on consumption, 65
  drawdown constraint on wealth, 40
  finite horizon, 30
  generic, 116
  habit formation, 35
  history-dependent preferences, 46
  labour income, 110
  leverage bound, 96
  limited look-ahead, 86
  Markov-modulated asset dynamics, 54, 56
  penalty for riskiness, 78
  production and consumption, 82
  random growth rate, 60
  random lifetime, 58
  recursive utility, 72
  retirement, 101
  soft wealth drawdown, 97
  stochastic volatility, 89
  stopping early, 68
  transaction costs, 37
  utility bounded below, 80
  utility from wealth and consumption, 62
  varying growth rate, 92
  Vasicek interest rate process, 31
  wealth preservation, 63

I
Implied volatility surface, 23
Inada conditions, 24
Infinite horizon, 3
Infinitesimal generator, 13
Innovations process, 55, 92
Insurance example, 49
Integration by parts, 18
Interest rate risk, 31
Interpolation, 124
Inverse marginal utility, 17

J
Jail, 79
Jones, keeping up with, 73
Jump intensity, 116
Jump intensity matrix, 53, 118

K
Kalman-Bucy filter, 92
Knaster–Kuratowski–Mazurkiewicz
    theorem, 26

L
Labour income, 110
Lagrange multiplier, 17
Lagrangian
  expected shortfall, 70
Lagrangian semimartingale, 18
Later selves, 86
Least concave majorant, 68, 121
Leverage bound, 96
Limited look-ahead, 84
Linear investment rule, 20
Log utility, 6

M
Marginal utility, 10
Market clearing, 23, 77
Market clock, 137
Market price of risk K, 9
Markov chain, 53
Markov chain approximation, 115, 122, 123
Markov-modulated asset dynamics, 53
Martingale principle of optimal
    control, 3, 65, 68
  transaction costs, 37
Merton consumption rate, γ_M, 9, 20
Merton portfolio π_M, 8
Merton problem, 1, 14
Merton problem, well posed, 20
Merton value, 9
Minimax, 107

N
Nash equilibrium, 83
Negative wealth, 79

Net supply, 24
Newton method, 135
  habit formation, 35
Non-CRRA utilities, 47
No-trade region, 38
Numerical solution, 115

O
Objective, 3
Offset process, 89
Optimal stopping, 120
Optional projection, 92
OU process, 59

P
Parabolic problems, 127
Parameter uncertainty, 102
Pasting, 38
PDE for dual value function, 12
Penalty for riskiness, 78
Policy improvement, 117, 134
  history-dependent preferences, 47
  Markov-modulated asset dynamics, 56
  random growth rate, 60
  transaction costs, 38
  Vasicek interest rate process, 32
Pontryagin-Lagrange approach, 17
Portfolio process, 2
Portfolio proportion, 3
Preferences
  history-dependent, 45
Production, 81
Production function, 81

Q
q-q plot, 139

R
Random growth rate, 59
Random lifetime, 57
Recursive utility, 72
Reflecting boundary conditions, 60
  Vasicek interest rate process, 32
Regime-switching model, 142
Representative agent, 25
Resolvent, 13, 27, 48
Resolvent density, 131
Retirement, 99
Riskless rate, 2
Robust optimization, 106

S
Scale function, 122
Scaling, 6
  annual tax accounting, 43
  drawdown constraint on consumption, 65
  drawdown constraint on wealth, 40
  finite horizon, 30
  habit formation, 34
  history-dependent preferences, 46
  Markov-modulated asset dynamics, 55
  production and consumption, 83
  random growth rate, 59
  random lifetime, 58
  transaction costs, 37
  varying growth rate, 92
  wealth preservation, 63
Slice of cake, utility from, 76
Soft wealth drawdown, 97
Standard objective, 29
Standard wealth dynamics, 29
State-price density, 10, 15, 19
  marginal utility, 22
State-price density process
  uncertain growth rate, 103
Static programming approach, 14
Stochastic optimal control, 115
Stochastic volatility, 88
Stochastic volatility model, 141
Stopping early, 68
Stopping sets, 120
Stylized facts, 139
Successive over-relaxation method, 117

T
Tax credit, 45
Time horizon, 3
Transaction costs, 36

U
Universal portfolio algorithm, 110
Utility bounded below, 79
Utility from wealth and consumption, 61

V
Value function, 4
Value improvement

V (*cont.*)
insurance example, 51
Value recursion, 134
Varying growth rate, 91
Vasicek interest rate process, 31
Verification, 10, 11, 17

W
Wealth equation, 2
Wealth preservation, 62
Wronskian, 131

MIX
Papier aus verantwortungsvollen Quellen
Paper from responsible sources
FSC® C105338

If you have any concerns about our products,
you can contact us on
ProductSafety@springernature.com

In case Publisher is established outside the EU,
the EU authorized representative is:
Springer Nature Customer Service Center GmbH
Europaplatz 3, 69115 Heidelberg, Germany

Printed by Libri Plureos GmbH
in Hamburg, Germany